ZERO DOWNTIME PNEUMATICS

Dr. Z. J. Lansky, P.E.
John J. Pippenger, P.E.

FPS
2433 N. Mayfair Road
Milwaukee, WI 53226

NOTICE

Information contained in this work has been obtained by FPS from sources believed to be reliable. However, neither FPS nor its authors guarantees the accuracy or completeness of any information published herein. Neither FPS nor its authors shall be responsible for any errors, omissions, or damages arising out of use of this information. This work is published with the understanding that FPS and its authors are supplying information, but are not attempting to render engineering or other professional services or advice. If such services are required, the assistance of an appropriate professional should be sought.

ISBN 0-929276-03-5
Printed in the United States of America

AMALGAM'S GLOBAL SERIES IN FLUID POWER

John J. Pippenger, P.E. Editorial Consultant

BASICS FOR THE FLUID POWER MECHANIC
Greg P. Gordon, P.E.
John J. Pippenger, P.E.

ZERO DOWNTIME PNEUMATICS
Dr. Z. J. Lansky, P.E.
John J. Pippenger, P.E.

MODERN HYDRAULICS: THE BASICS AT WORK
Dr. William D. Wolansky
Dr. Arthur Akers

Authored by John J. Pippenger, P.E.

FLUID POWER: THE HIDDEN GIANT

HYDRAULIC CARTRIDGE VALVE TECHNOLOGY

ZERO DOWNTIME HYDRAULICS

ACKNOWLEDGEMENTS

Proprietary educational materials have been very effective in guiding the mechanics, designers and technicians in the use and maintenance of pneumatic power transmission systems.

Our expanding technology-based society is becoming increasingly dependent upon versatile and dependable power transmission systems appropriate to our changing life style.

Pneumatic and vacuum-powered machinery offers clean, versatile, dependable and precise force and motion. Thus the machinery of today depends heavily on this technology.

This generic text has been prepared for entry level mechanics and machine operators who are seeking information to acquaint them with this versatile power transmission option. We believe it will also be of significant use to the technician and designer of industrial machinery.

An especial thanks goes to Gunner Grotness of PHD, Inc., Fort Wayne, Indiana who arranged a round robin review of the initial manuscript by members of the National Fluid Power Association Pneumatic Council. The comments and suggestions were very much appreciated with an obvious contribution to the contents of this manuscript.

We also appreciate the contribution of illustrations, charts and technical information by the manufacturers of pneumatic systems and components. Their follow-up review of the manuscript for technical and educational content was most helpful.

Similar assistance was provided by members of the FPS. It is expected that this guidance will be of assistance to those mechanics, technicians and engineers seeking certification in one or more of the Fluid Power Society programs.

Readers interested in the Fluid Power Certification programs can obtain further information by contacting the FPCB Secretariat, c/o FPS at 2433 N. Mayfair Road, Suite 111, Milwaukee, Wisconsin 53226. Telephone: (414) 257-0910 Fax: (414) 257-4092.

Dr. Z. J. Lansky, P.E.

John J. Pippenger, P.E.

PREFACE

The fragmented educational channels presenting fluid power applications in the real world have pinpointed the need for user-friendly communications.

The acceptance of my book Zero Downtime Hydraulics by the designers, distributors and users of hydraulic power transmission systems suggested that a similar book addressing power transmission via compressed air might serve a comparable purpose for the designers and users of pneumatic systems.

We, as authors, can point out areas where our mistakes over many years of practical experience need not be repeated. Much of our personal experience has been supplemented by articles presented at meetings of the National Conference on Fluid Power and magazine articles from a host of technical publications.

An excellent presentation by Frank Yeaple in a Design News supplement supplied significant current information garnered from suppliers within the pneumatic industry. The National Fluid Power Association Pneumatics Council's support for and development of friendly communications also aided in structuring this manuscript.

The Fluid Power Distributors Association has continually supported an educational program which includes significant pneumatic content.

The distributors have been the backbone of pneumatic design information serving as an effective communication channel between the manufacturer and user of pneumatic systems.

Our objective here is to acquaint the entry level mechanic with pneumatic power, remind the designer of some of the basics that are often overlooked, supplement the texts used in technical training programs and to suggest that a considerable amount of technical assistance is available through the efforts of the many experts in the distribution network who are selling pneumatic products and systems. Our objectives will have been accomplished if we have helped you eliminate unnecessary downtime. We believe this will happen.

To quote Frank Yeaple - "What other power and control medium is as plentiful as air? As inexpensive? As resistant to temperature extremes? As immune to short circuits? As free of oil leaks? As easy to work with?" Frank concludes "The answer is none! Pneumatic power is definitely unique".

Dr. Z. J. Lansky, P.E.

John J. Pippenger, P.E.

CONTENTS

ACKNOWLEDGEMENTS **V**
PREFACE **VII**
CONTENTS **IX**
1. INTRODUCTION **1**
STEAM COULD PROVIDE POWER
WITH LOCAL COMBUSTIBLES 4
EFFICIENT COMPRESSED AIR SYSTEMS
REQUIRE GOOD MACHINING 4
ACADEMIC CHANNELS DEVELOP AT A SLOW PACE 4
THE ELECTRIC MOTOR REVOLUTIONIZED
POWER DISTRIBUTION 4
HYDRAULIC POWER TRANSMISSION SYSTEMS
ARE USUALLY SELF-CONTAINED 6
PNEUMATIC POWER TRANSMISSION SYSTEMS
FALL INTO TWO CATEGORIES 6
SUMMARY 9
REVIEW QUESTIONS 9
2. OPERATOR AND SERVICE
PERSONNEL TRAINING **11**
CIRCUIT SYMBOLS - MEASURING FLOW
AND PRESSURE 11
FLUID POWER SYMBOLS 13
CIRCUIT ELEMENTS 14

QUICK CONNECTIONS 16
FLUID POWER COMPRESSOR AND MOTOR SYMBOLS 17
HYDRAULIC CYLINDERS (LINEAR MOTORS) 18
SYMBOLS FOR DIRECTIONAL VALVES 20
CONDITIONING OF THE AIR 23
SUMMARY 26
REVIEW QUESTIONS 26
3. THE GAS LAWS 27
THE AIR AROUND US 27
THE FIRST BAROMETER 28
SEA LEVEL WAS A GOOD STARTING POINT 30
THE GASES THAT MAKE UP OUR ATMOSPHERE 30
DRY AIR AT SEA LEVEL 32
STANDARD AIR AND FREE AIR 35
A REVIEW AND THE APPLICATIONS OF BOYLE'S LAW 39
THE PNEUMATIC SPRING 39
AIRCRAFT ACCUMULATORS 41
ADDING GAS CAPACITY 41
TEMPERATURE CONSIDERATIONS 41
THE GAS LAWS 42
PASCAL'S CONCLUSION 42
BERNOULLI'S PRINCIPLE 42
PNEUMATIC EJECTION SYSTEMS 43
SUMMARY 43
REVIEW QUESTIONS 43
4. FORCE AND MOTION 45
INITIATION AND CONTROL OF FORCE AND MOTION 45
THE POWER FUNCTION - THE MOTION FUNCTION 45
THE POWER FUNCTION 45
THE MOTION FUNCTION 46
THE PNEUMATIC ADVANTAGE 46
THE PNEUMATIC DISADVANTAGE OR LIMITATION 48
SUMMARY 49
REVIEW QUESTIONS 49
5. THE POINT OF USAGE 51
MOVING MATERIALS 51
CAPTURING THE POWER 51
ENJOYING THE POWER 55
REVERSING THE FLOW 55
THE POWER ASSOCIATED WITH INFLATION 59
THE POWER ASSOCIATED WITH DEFLATION 62
POWER TRANSMISSION VIA CONTAINED AIR 62
FLOATING ON AIR 65

LINEAR FORCE AND MOTION 65
BASIC CYLINDER DESIGN 66
CONSTRUCTION FEATURES SHOULD BE
CAREFULLY CONSIDERED 66
PRESSURE RATINGS 67
WORKING MEDIUM 69
ASSEMBLY OF TIE ROD-TYPE CYLINDER IS CRUCIAL 69
CYLINDER MOUNTING TYPES 70
MOUNTING SPECIFICATIONS 70
FORCES IN TENSION OR COMPRESSION 73
POTENTIAL ALIGNMENT PROBLEMS 73
NOTES ON INSTALLATION 73
ROD END JOINTS 77
PISTON ROD DIAMETER 78
CIRCUIT CONSIDERATIONS 79
CUSHIONING 80
ROD SEALS 82
STROKE SPEED CONTROL 84
ROTARY FORCE AND MOTION 86
VERSATILE PNEUMATIC ACTUATORS 88
THE CASE OF THE SINGING CYLINDERS! 92
SUMMARY 96
REVIEW QUESTIONS 96

6. COMPRESSING AIR - THE AIR PUMP 97

COMPRESSSED AIR SERVED THE ANCIENTS 97
THE INDUSTRIAL REVOLUTION
PROVIDES REFINEMENTS 98
BASIC COMPRESSOR OPERATION 100
MINIMIZING OIL IN THE OUTLET LINES 102
THE SLIDING VANE COMPRESSOR 103
NON-POSITIVE DISPLACEMENT-TYPE
COMPRESSORS 106
UNIT COMPRESSORS 107
PORTABLE COMPRESSORS 107
SUMMARY 109
REVIEW QUESTIONS 110

7. EXHAUSTING AIR - THE VACUUM PUMP 111

VACUUM IS ESSENTIAL THROUGHOUT
OUR LIFE SPAN 111
THE MECHANICAL VACUUM PUMP LOOKS
LIKE AN AIR COMPRESSOR 111

VACUUM IS ESSENTIAL TO THE HOUSEKEEPER 112
INDUSTRY USES VACUUM FOR BOTH
MATERIAL AND PRODUCT MOVEMENT 112
VACUUM PUMPS ARE TAILORED TO
SPECIFIC NEEDS 113
SOME TYPICAL APPLICATIONS OF
VACUUM IN INDUSTRY: 114
1. VACUUM DOES NOT MOVE MATERIAL
BUT AIR VELOCITY DOES 114
2. HOW THE SUCTION CUP WORKS 115
3. TYPICAL APPLICATIONS OF THE
SUCTION CUP 118
4. VACUUM SUCTION NOZZLE 118
5. VACUUM SUCTION HEAD 118
6. VACUUM FORMING 118
THE MULTI-EJECTOR CONCEPT 120
FILTERS TO PROTECT
THE VACUUM PUMP AND ACCESSORIES 122
SUMMARY 124
REVIEW QUESTIONS 124
8. THE STORAGE FUNCTION 125
THE AIR RECEIVER 125
SIZING A COMPRESSED AIR RECEIVER 125
RECEIVER DESIGN WITH ON-OFF REGULATION 127
DUAL PRESSURE PNEUMATIC SYSTEM 127
AUTOMATIC ELECTRIC DRAIN VALVE 127
SUMMARY 130
REVIEW QUESTIONS 130
9. THE DISTRIBUTION SYSTEM 131
INITIAL SYSTEM DESIGN
CAN MINIMIZE FUTURE PROBLEMS 131
A LOOK TO THE FUTURE 131
SELECTING PIPE DIMENSIONS 133
PIPING AND INSTALLATION OF AUXILIARIES 133
TRAPPING THE WATER AND
COMPRESSOR LUBRICANTS 136
LOW LEVEL INSTALLATION FOR ACCESSIBILITY 136
HIGH LEVEL INSTALLATION HAS
SOME ADVANTAGES 137
PIPING AND AUXILIARIES FOR MACHINE TOOLS 137
LINES BETWEEN CONTROL AND ACTUATOR 138
SUMMARY 139
REVIEW QUESTIONS 139

10. CONDITIONING THE AIR **141**
PROPER CONDITIONING OF AIR
IS A PRELUDE TO GOOD PERFORMANCE 141
AIR-LINE FILTERS 142
FILTER ELEMENTS 142
CLOTH OR FELT-TYPE ELEMENTS 142
HIGH FILTRATION WITH MINIMUM
PRESSURE DROP 143
ADVANTAGES AND DISADVANTAGES 143
WIRE-MESH ELEMENTS 143
SINTERED-BRONZE OR POROUS STONE 143
LIQUID REMOVAL 144
TANK DRAINS, SEPARATORS AND DRYERS 145
MOISTURE SEPARATORS 145
INSTRUMENT PURE AIR 145
DESICCANTS 146
TANK DRAINS 148
SEPARATORS 148
AIR DRYERS 149
AIR LINE LUBRICATORS 150
INSTALLATION OF LUBRICATORS 151
POTENTIAL START-UP PROBLEMS 151
LUBRICATION FOR AN AIR CYLINDER 151
KEY LUBRICATION POINTS TO REMEMBER 153
CHARACTERISTICS OF LUBRICATORS 153
HOW MUCH LUBRICANT? 155
LUBRICATOR OIL 155
LUBRICATOR SYSTEMS 156
COMBINED FILTER-REGULATOR-LUBRICATOR UNITS 156
SUMMARY 156
REVIEW QUESTIONS 157
11. HEAT TRANSFER **159**
DISSIPATING HEAT OF COMPRESSION 159
COMPRESSOR AIR 160
THE INCREASE IN POTENTIAL ENERGY 160
THE FUNCTION OF THE AFTERCOOLER 160
AN AFTERCOOLER CONDENSES WATER VAPOR 160
OVER COMPRESSION 162
REFRIGERATION 162
SUMMARY 163
REVIEW QUESTIONS 163
12. EXHAUST CONSIDERATIONS **165**
COMPLETING THE POWER CYCLE 165

THE REFRIGERATING POTENTIAL 165
LUBRICATION DISPERSION 166
THE MUFFLER 167
SUMMARY 169
REVIEW QUESTIONS 169
13. AIR LINES AND CONNECTORS 171
FLEXIBLE PLASTIC TUBING AND
QUICK CONNECT/DISCONNECT FITTINGS 171
FLEXIBLE PLASTIC TUBING 172
PUSH-IN TYPE FITTINGS 173
1. SPRING LEVERED TOOTH 174
2. BRASS COLLET 174
3. METAL RING SQUARE TOOTH 175
4. PLASTIC COLLET METAL RIDGE 175
5. PLASTIC RETAINER 175
INSTALLATION AIDS FOR PLASTIC TUBE
AND ASSOCIATED FITTINGS 175
IF POSSIBLE AVOID PIPE FITTINGS 176
TUBE SUPPORT 180
TUBE CUTTING WITH A HACKSAW 183
TUBE DEBURRING 185
DEBURRING TOOLS 185
THREAD PROTECTORS 186
TUBE INSTALLATION FOR CONVENIENT SERVICING 187
HOSE AND TUBING OPTIONS 187
KEEP LENGTH OF AIR HOSE AS SHORT AS POSSIBLE 188
OPTIONAL LINE MATERIALS 189
HOSE AND TUBE FITTINGS 189
QUICK-DISCONNECT COUPLINGS 190
ROTARY AND SWIVEL JOINTS 194
SUMMARY 194
REVIEW QUESTIONS 195
14. CONTROL VALVING 197
PRESSURE CONTROL VALVES 198
PILOT-OPERATED REGULATOR 201
DIGITAL AIR PRESSURE CONTROL WITH
A SOLENOID PILOT ASSEMBLY/REGULATOR 202
OPERATING CHARACTERISTICS 205
CHOICE OF REGULATOR 205
DIRECTIONAL CONTROL VALVES 206
SHEAR-ACTION DESIGN 206
PISTON OR SPOOL VALVE 206
PLUNGER VALVE 208
DISK VALVE 208

PLUG VALVE 208
SEATING-ACTION DESIGN 208
POPPET VALVE 210
DIAPHRAGM VALVE 210
VALVE PORTING 210
TWO-WAY VALVES 210
THREE-WAY VALVES 211
FOUR-WAY VALVES 211
FIVE-WAY VALVES 211
VALVE OPERATORS 212
MANUAL OPERATORS 212
MECHANICAL OPERATORS 213
SOLENOID OPERATORS 215
PILOT OPERATORS 218
DIRECT-PILOT OPERATORS 218
LOW-PRESSURE PILOT OPERATORS 218
BLEED-PILOT OPERATORS 219
DIFFERENTIAL- PILOT OPERATORS 220
RETURN OPERATORS 221
DETENTS 221
VALVE POSITIONS 221
TWO-POSITION 221
THREE-POSITION 221
NORMALLY-OPEN 221
NORMALLY-CLOSED 223
FLOW CONTROL VALVES 223
FLOW-CONTROL, SPEED-CONTROL, AND BLEED-CONTROL VALVES 223
QUICK-EXHAUST VALVES 225
CHECK VALVES 226
CUTOUT VALVES 227
SELECTOR VALVES 227
SEQUENCE VALVES 227
SHUTTLE VALVES 227
SUMMARY 228
FOR ZERO DOWNTIME: 228
REVIEW QUESTIONS 228

15. MANIFOLDS AND SUB-BASES 231

MOUNTING METHODS AND ARRANGEMENTS 231
DIRECT-MOUNTING 231
SUB-BASE MOUNTING 232
STACKING AND MANIFOLD MOUNTING 235
SUMMARY 235
REVIEW QUESTIONS 237

16. LOGIC CIRCUITRY **239**
FLUIDICS 240
THE MICROPROCESSOR 240
MULTI-TALENTED DESIGN AND SERVICE 242
SUMMARY 242
REVIEW QUESTIONS 243
17. SAFETY CONSIDERATIONS **245**
1. TRAINING THE OPERATOR 245
2. TRAINING SERVICE PERSONNEL 246
3. TRAINING MAINTENANCE PERSONNEL 247
4. THE MACHINE DESIGNER'S RESPONSIBILITY 247
5. THE INSTALLERS RESPONSIBILITY 249
MINIMUM TRAINING REQUIREMENTS: 250
OPERATOR 250
SERVICE PERSONNEL 250
SET-UP PERSONNEL 250
MAINTENANCE PERSONNEL 250
MACHINE DESIGNER 251
INSTALLATION CREW 251
SUMMARY 251
REVIEW QUESTIONS 252

1

INTRODUCTION

History tells us very little about the early uses of induced air movement. Little imagination is needed to recognize the early use of the human lungs to assist in the fire-making chore (Figure 1.1). The blowgun, powered by the human lungs offered a faster and more accurate method of hunting the fleet prey (Figure 1.2). These simple applications of compressed and/or moving air depended on the human lungs as the prime power source.

Early power-driven devices used several different energy sources. History records the use of wind power for many types of sailing ships plus the versatile windmill (Figure 1.3). A second source recorded throughout history is water power. This could be flow forces from the liquid in motion down a stream or the weight of the falling water providing the power (Figure 1.4). Waterways, sluices, and/or troughs could direct water. The Roman aqueducts were well known at the time of Christ. Some water-directing systems had agricultural irrigation as a primary objective and power recovery as a secondary objective. Wheels could be turned by the velocity of the stream or by falling water with the weight providing the power. The Pelton Water Wheel captured greater energy by selectively directing the flow against bucket wheel assemblies (Figure 1.5). The vertical water driven turbine is widely used in hydroelectric power plants.

Figure 1.1 Moving air assists in the fire-making chore.

Figure 1.2 The air-propelled dart was often faster than the clumsy rock.

Figure 1.3 The windmill can pump water, grind grain and do other heavy chores with the natural air movement as the power source.

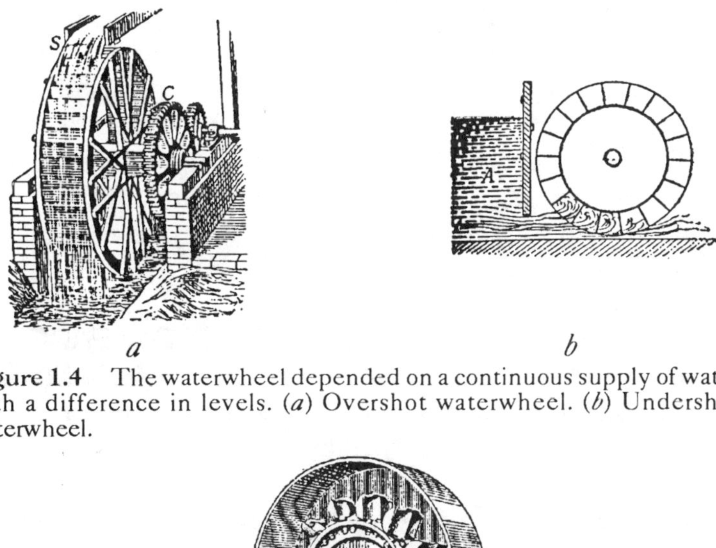

a *b*

Figure 1.4 The waterwheel depended on a continuous supply of water with a difference in levels. (*a*) Overshot waterwheel. (*b*) Undershot waterwheel.

Figure 1.5 The Pelton Water Wheel greatly improved the energy recovery from the moving water.

STEAM COULD PROVIDE POWER WITH LOCAL COMBUSTIBLES

Practical use of steam as a power source expedited the technological growth of the eighteenth and nineteenth century. Local supplies of wood, coal, oils and other combustibles fed the steam boiler (Figure 1.6).

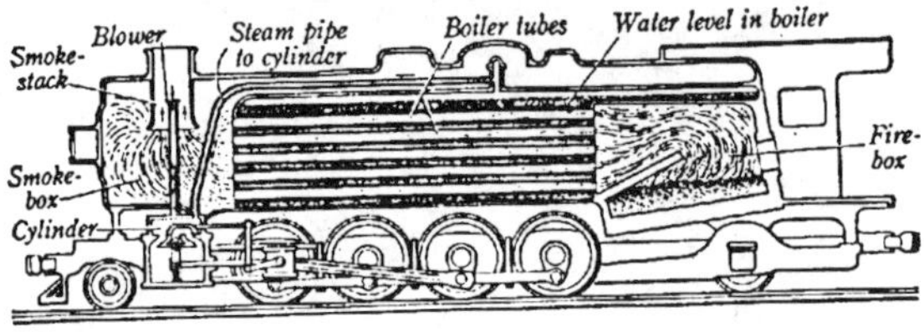

Figure 1.6 The locomotive converted locally available combustibles into motive power.

EFFICIENT COMPRESSED AIR SYSTEMS REQUIRE GOOD MACHINING

In 1670 Papin was able to compress air with a water wheel. It took many more years of development before the pneumatic power potential was economically feasible. The natural dynamics of air movement certainly provided power at an early period in history. The compressing and storing of air was dependent upon the development of efficient cylinders, gear pumps, and the piping to channel the compressed air from the compressor mechanism to the actuator or work device.

ACADEMIC CHANNELS DEVELOP AT A SLOW PACE

Historically, the classical academic study of power transmission has concentrated primarily on water hydraulics and in such mechanical areas as gears, levers, belts, etc. The early study of steam devices may be an exception.

Of course, early systems were mostly water. The objective was to get power out of the moving water where there was a sudden change of height, such as with a natural waterfall. Ponds at different levels were created to raise the water level and store it by the development of dams. In some areas the ocean tide levels could be harnessed for power.

THE ELECTRIC MOTOR REVOLUTIONIZED POWER DISTRIBUTION

With the advent of electric motors (both direct current and alternating current types) another power transmission option was introduced into the academic channels with significant emphasis and degree status. Fluid power, another basic option for power transmission, has not attained the broad academic stature accorded to mechanical, steam and electrical systems.

Industrial fluid power is not taught in many academic degree programs, but those studies currently available do cover fluid mechanics and associated engineering curricula. The broad subject of fluid mechanics is available in virtually every major academic technological institution.

Movement and characteristics of fluids, both liquids and gasses are well documented. The use of a suitably contained and pressurized fluid to transmit power in linear or rotary mode has had modest exposure in the academic community.

Major hydraulic power transmission systems emerged early in the Industrial Revolution, closely associated with the development of steam power technology and the steam engine. Both of these power transmission systems depended completely on the ability to machine round bores, pistons and rods with surfaces smooth enough to tolerate a resilient sealing mechanism so as to contain the pressurized liquid or gas without excessive friction and/or abrasion.

The availability of the piston structure which could be easily sealed was essential to both steam and water-type hydraulic power systems, and ultimately to pneumatic power systems. Pneumatic systems have also used bellows and diaphragm devices in the power transmission chain.

The early hydraulic power systems were typically associated with linear force and motion. The use of a gas under pressure to transmit power also used many of the basic structures associated with steam power transmission systems. Pneumatic power transmission systems currently available have many roots in steam power transmission technology.

The early central systems associated with steam, water hydraulic power systems and pneumatic power transmission systems have remained an important part of many industrial pneumatic systems. Many modern submarines and power plants are steam-driven with the steam generated from the heat of the nuclear reactor. What has been replaced is the burning oil or other heat source coal, geothermal heat, etc.

Steam, as a power transmission system, has been relegated to some important specialized areas. Heat engines, using internal combustion, diesel and turbine power have decreased the broad use of steam for power transmission. Low cost electric power has also been a significant factor in the demise of the broad use of steam power, recognizing of course, that steam driving a turbine is a key basic power chain element in the generation of electricity.

HYDRAULIC POWER TRANSMISSION SYSTEMS ARE USUALLY SELF-CONTAINED

Hydraulic power transmission systems (with a few exceptions) are self-contained units with the pump(s), actuator(s) and controls specifically designed for predetermined power transmission tasks.

PNEUMATIC POWER TRANSMISSION SYSTEMS FALL INTO TWO CATEGORIES

Pneumatic power transmission systems fall into two broad categories: central systems and portable units. Central systems employ one or more air compressors delivering compressed air into a common reservoir. This provides air under pressure to supply the lines leading from the receiver tank to many points of application.

At the point of usage it is good practice to provide a filter, lubricator and regulator. The air from the receiver should have been cooled and filtered to a specific level and delivered to the application point at a chosen maximum pressure potential. The pressure potential is limited to the supply value from the compressor(s) or some lesser pressure as determined by the adjustment of a branch regulator and the effects of line restrictions and friction. Virtually all pneumatic systems are structured so that air used at the point of application is returned to the atmosphere. Return exhaust air lines are rarely used. Under some circumstances it may be desirable to direct the exhaust air to a specific area to avoid contamination or operator ingestion of entrained lubricants.

Virtually all air compressors deliver air to some kind of receiving tank. The exception would be those used for constant flow devices or small, self-contained circuits such as shaking tables, driving rotary air devices, blower service and other very specific applications.

Pneumatic systems often have multiple point-of-use functions. In addition to basic power transmission by use of linear cylinders or rotary motors, the pressurized air supply may be used to inflate tires and other devices, blow chips, clean molds and other cavities, cool exposed surfaces and supply power to spray paint. The compressed air may be used to transport parts – propel as well as support – by the use of "ground effect" machines in the movement of the heavy loads (Figure 1.7). In automobile service stations compressed air is used to move grease and other lubricants. The movement of wet or dry products may be in a planned conveyor mode (Figure 1.8).

Figure 1.7 Compressed air can be used to move heavy loads at high speed over selective terrain or water by creating a pressurized pad under the device to be moved. The Hovercraft is an excellent example.

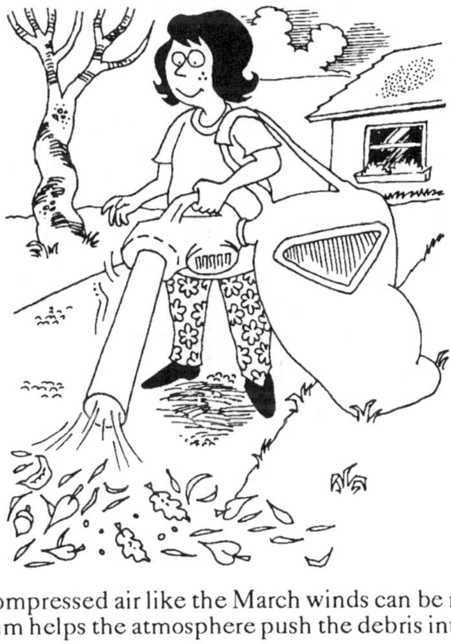

Figure 1.8 Compressed air like the March winds can be nature's broom. Creating a vacuum helps the atmosphere push the debris into a porous bag.

Some end use applications will use branch lines from the principal supply equipped with a regulator, filter and lubricator. This is further protected by a pocket in the pipeline where moisture can be trapped and removed prior to entering the work area through the filter, lubricator and regulator. Most air line filters are reasonably good at removing water and liquid oil carried by the flowing air. The device needs to be drained regularly. An automatic drain is preferable.

To further protect the branch circuits the take-off connection to the air system is made at the *top* of a supply line. Thus moisture can be retained in the supply line, collected in a trapping device, and expelled from the line to minimize formation of rust and scale. Reservoirs are also used to store air close to the point of usage in circuits where the demand can swing widely.

Portable pneumatic air compressor units are produced in tiny, fractional horsepower, specialized units for instrumentation and similar applications to giant wheeled units used for construction projects and demolition tools.

Portable pneumatic air compressor units (usually very large with significant capacity) may be used to supplement central systems when emergency or intermittent additional capacity is needed. If such usage is planned for, means of quickly attaching the supplementary unit and isolating the normal air supply device should be provided.

Vacuum systems can be considered as an inverse of the air compressor. Many vacuum systems are dedicated to material movement, such as the home vacuum cleaner. Particulate dirt and foreign matter is moved from carpet surfaces to a suitable container for disposal purposes. Many industrial vacuum systems serve in a similar mode. Our interest is in vacuum devices used as a power transmission system.

Vacuum-forming of plastics and sheet metal serves many industries. Vacuum cups employed in robotic machinery are widely used to pick-up, hold during a movement, and position even fragile materials such as sheet glass, metals and plastics in assembly, processing and packaging systems.

Alternate use of vacuum and compressed air may be programmed in a circuit employing a temporary resilient cup-type fastener. The vacuum applied to this gripping device will fasten it to the part to be handled. A shift to compressed air quickly releases the part.

Appropriate maintenance and operation within the design parameters after the installation procedure can be the basis for attaining *Zero Downtime Pneumatics*.

SUMMARY

Contained air compressed to a specific pressure level provides a convenient and efficient method of storing energy. The energy can be easily converted to force and motion, which can be in a linear and/or rotary pattern.

Power transmission systems employing vacuum can be considered an inversion of the compressed air mode with significant advantages in plastic and metal product forming and material handling. Vacuum is widely used in the printing trades for paper handling.

Compressed air and vacuum systems are inherently clean with minimum risk of product contamination. Pneumatic systems can be employed in areas subject to dust, explosion and electrical disturbances.

REVIEW QUESTIONS

1. What are the two broad categories into which pneumatic transmission systems fall?
2. At the point of usage it is considered good practice to provide three basic units. What are they called?
3. Where should the takeoff connection to a branch line be located?
4. What is the advantage of vacuum forming?
5. When did compressed air systems become significant in industry?
6. How does pneumatics relate to steam and hydraulics?
7. What is the major difference between hydraulic and pneumatic systems?
8. How is vacuum used in the manufacturing processes?
9. How is moisture removed from compressed air?
10. What is the purpose of a receiving tank?

2

OPERATOR AND SERVICE PERSONNEL TRAINING

CIRCUIT SYMBOLS - MEASURING FLOW AND PRESSURE

Communications provide an essential key to Zero Downtime Pneumatics. Certain key reactions to a positive action such as pushing a button, pulling or twisting a lever can provide visual cause and effect patterns.

The individual initiating the action has learned some of the basics of the machine functions. Unfortunately this is the extent of the training for many operators (Figure 2.1). Some individuals responsible for the health and welfare of the machine have little or no knowledge base of it beyond the cause and effect function. In a hydraulic system the flow of a few cc's can drop pressure to zero. In a pneumatic system there can be a significant supply of pressurized air which will maintain pressure even though there are internal or external leaks.

A knowledge of symbology, especially the symbols for pneumatic, electric and logic circuit devices is a key to understanding circuitry controlling the force and motion functions of any machine employing a pneumatic power transmission system.

Symbols incorporated in a machine circuit drawing illustrate the pressure level control functions and the mechanisms used for the control of direction and rate of flow. Symbols also show the energy source and functions of actuators plus fluid conditioning devices and all devices affecting the operation of the pneumatic system. The following section on

Figure 2.1 Operator training may be limited to that needed for a specific production routine. (Courtesy of Ross Operating Valve Company)

fluid power symbols will acquaint you with some of the symbols used for both hydraulic and pneumatic circuits. Some of the symbols are used exclusively in pneumatic circuits, some are used exclusively for hydraulic components and circuits. As an example, virtually all pneumatic systems pressurize free air from the atmosphere surrounding the compressor. At the completion of the pneumatic force and motion functions the air is again returned to the atmosphere. An open circuit hydraulic system employs a reservoir to contain the liquid to be moved and pressurized. This liquid is pumped from the reservoir, through the circuitry to provide the desired force and motion functions and then returned to the reservoir. A closed circuit hydraulic system employs a reversible pump which provides direction and rate of flow for the actuator whether it be a motor for rotary force and motion or a cylinder for linear force and motion. Pressure level is a function of the load encountered by the actuator.

Maximum pressure level can be established with appropriate relief valve(s) which divert the fluid to a lower pressure area upon reaching the set maximum system pressure. Because the pump and actuator of the closed hydraulic system must have lubrication and cooling, the fluid used in the high pressure power transmission loop is supplemented by an auxiliary supply from a charge or makeup pump which is a key component in the hydrostatic transmission. The makeup pump operates at a relatively low pressure, supplying the low pressure portion of the hydrostatic circuit with fluid to replace that lost in the leakage, lubrication and cooling functions primarily supplied from the high pressure side of the power system. A small reservoir is included in the hydrostatic circuit to supply the makeup pump. The makeup pump output replaces the lost fluid in the hydrostatic circuit and usually conditions the fluid by filtration and cooling before returning it to the power circuit.

Thus it is obvious that the symbology employed in a pneumatic system may be similar to that used in a hydraulic system. The receiving tank employed in a pneumatic system may be loosely compared to the reservoir in the hydraulic system with the recognition that the holding tank is always pressurized in a functional pneumatic circuit and the hydraulic reservoir may or may not be pressurized.

FLUID POWER SYMBOLS

The introduction and development of fluid power symbols through the efforts of the Joint Industry Conference (JIC) of the 1940's and subsequent activities by members of the American Society of Mechanical Engineers (ASME), ANSI (American National Standards Institute), ISO 1219 (International Standards Organization) and others must be considered a monumental step in the development of communication channels for the fluid power industry.

Graphic communications, like languages, are easiest to learn if they can be directly identified with the associated subjects. This presentation acquaints you with the basic symbols, terminology and rules associated with the use of graphic symbology for fluid power systems. The specific use of the symbols will be expanded as you continue your study of fluid power components and circuits which is essential to attain troubleshooting skills. A working knowledge of the material in this chapter will simplify your work in relationship to any fluid power system. It is especially valuable in those systems involved with complete manifold assemblies. The symbols are the most practical and easiest route to follow the flow and power pattern through the power transmission system.

In summary - hydraulic symbols are used throughout the world in the design, operation and maintenance of fluid power systems. A knowledge of both pneumatic and hydraulic symbols will enable you to read and understand circuit diagrams and other important drawings of fluid power systems. This knowledge is a valuable aid in your work with all types of pneumatic equipment. Some of the symbols presented in this section are called *simplified symbols*. Other more complex symbols will be explained as you study specific components (valves, controls, etc.) in later chapters. Simplified symbols are widely used in commercial circuit diagrams. More complex symbols may be used when a basic valve or control is modified to make it perform a new or different function.

CIRCUIT ELEMENTS

A fluid power circuit consists of two or more fluid power devices connected by piping or tubing. Connections within a manifold assembly may be via drilled, cast or machined passages. In a diagram or drawing of a fluid power circuit, a solid line (Figure 2.2*a*) represents a *pipe*, a *tube*, or some *other conductor* capable of handling major flow in the circuit appropriate to the rate of flow and pressure level to be contained. This type of pipe is often termed a *major conductor*. It conveys fluid to various parts of the circuit.

Pilot lines are usually of much smaller fluid-carrying capacity (and diameter) than major lines. Figure 2.2*b* shows a pilot line. The length of each dash is at least 20 line-widths, with the space between dashes approximately 5 line widths. The fluid pressure in pilot lines is usually the same as that within the major piping. Pilot-line pressure may, however, be any value needed in the circuit.

Fluid-drain and *air-exhaust lines* (Figure 2.2*c*) may be made of lighter weight material, not able to withstand as high a pressure as pilot or major pipes. The length of the dashed lines and the space between them in the symbol are about equal, each being approximately 5 line-widths.

Pipes that cross each other (but are not interconnected) are drawn with a small loop (Figure 2.2*d*). Note that Figure 2.2*d* shows two *major lines* (or pipes) crossing, but not interconnecting. Figure 2.2*e* shows a major line being crossed by a pilot line (or pipe). In fluid power terminology the words *pipe*, *tubing*, *line* and *conductor* are often used interchangeably to designate the structure that conveys the fluid from one device to another. Figure 2.2*f* shows two drain lines crossing each other without physical interconnection. Figure 2.2*g* shows two major lines crossing without the loop. This will occasionally be encountered in some industries, particularly processing plants. It is not a preferred drawing practice for fluid power circuits.

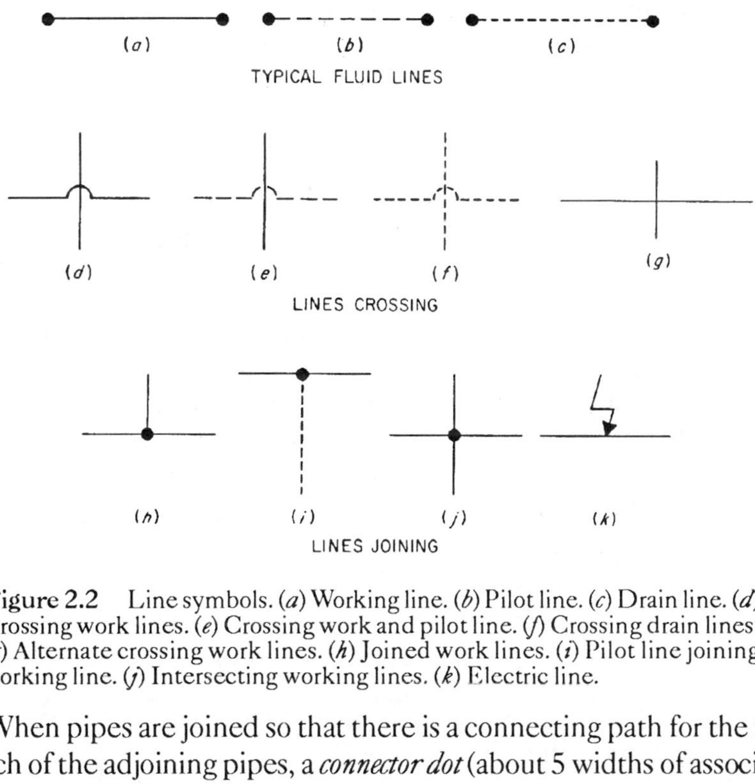

Figure 2.2 Line symbols. (*a*) Working line. (*b*) Pilot line. (*c*) Drain line. (*d*) Crossing work lines. (*e*) Crossing work and pilot line. (*f*) Crossing drain lines. (*g*) Alternate crossing work lines. (*h*) Joined work lines. (*i*) Pilot line joining working line. (*j*) Intersecting working lines. (*k*) Electric line.

When pipes are joined so that there is a connecting path for the fluid to each of the adjoining pipes, a *connector dot* (about 5 widths of associated symbol lines) is placed on the diagram to clearly show the point of junction. This is a preferred practice for fluid power circuits. Figure 2.2*h* shows two major lines joined at the point designated by the dot. In an actual circuit, this is usually a pipe or tubing tee. Figure 2.2*i* shows the point at which a pilot line is connected to a major line. The pilot line is usually smaller in diameter, and the connection may be a tee with a reduced-diameter outlet to the pilot line. Figure 2.2*j* shows two major lines intersecting. The dot indicates that these lines are interconnected. A fitting in the form of a *cross* with four connections or two tees in close proximity may be used for this connection. Note that these symbols do not indicate whether the piping is fabricated from *tube and fittings* or from other types of material. Instead, the symbols show the path of the fluid through the circuit. Electric lines are shown by Figure 2.2*k*. Figure 2.3*a* shows how a flexible connection is indicated in a circuit diagram. This symbol indicates some type of hose or similar flexible device.

An equilateral triangle (Figure 2.3*b*) is placed on a fluid line to show the direction in which the fluid is expected to travel. If a solid equilateral triangle is used, it indicates that the fluid is a liquid. An outline of an

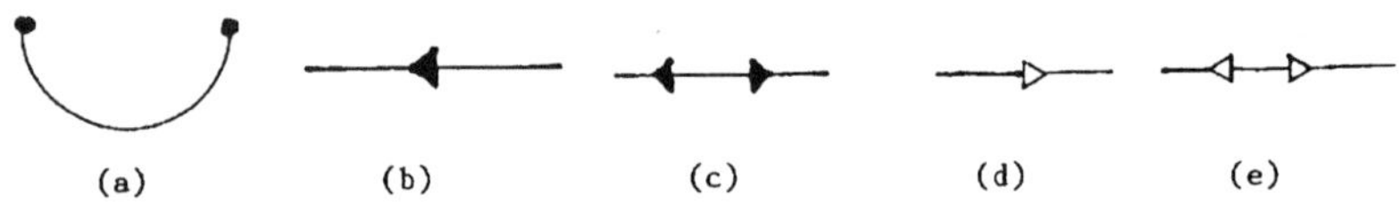

Figure 2.3 (*a*) Flexible line. (*b*) Energy flow (hydraulic) in one direction only. (*c*) Energy flow (hydraulic) in both directions. (*d*) Energy flow (pneumatic) in one direction only. (*e*) Energy flow (pneumatic) in both directions.

equilateral triangle, such as Figure 2.3*d*, indicates that the fluid is compressed air or a gas. If the fluid can travel in both directions at different times, two equilateral triangles facing in opposite directions are used (Figure 2.3*c* and 2.3*e*). Flow-direction triangles can be omitted from circuit diagrams unless there is a definite need for an indication of the flow direction or the fluid medium. Hydraulic and pneumatic nozzles are indicated by the symbol shown in Figure 2.4.

Figure 2.4 Nozzles.

QUICK CONNECTIONS

When it is desired to connect two pipes, or more often a pipe and a hose to a hydraulic unit quickly, *quick connect couplings* may be used. These couplings employ mechanical locks that are easily released and reassembled. Flexible hose of some type is generally used with the coupling. Tools are not needed to disconnect or reconnect. Some couplings are fitted with integral check valves that close when the coupling is disconnected. Closing of the check valve when the two halves of the quick coupling are separated prevents loss of air from the unit served by the coupling.

Figure 2.5*a* shows two quick-disconnects coupled together without integral check valves. One-half of a quick-disconnect coupling with a check valve to prevent fluid loss is shown in Figure 2.5*b*. Figure 2.5*c* shows a connected coupling with half equipped with a check valve and the other half without a check valve. Figure 2.5*d* shows the symbol for coupled connectors with check valves in both halves. Many quick connects used with agricultural implements and construction machinery tools have protective covers which slide into place as the units are disconnected. Some have caps attached with chain to be assembled to the open ends as the connectors are disconnected.

Figure 2.5 (*a*) Quick-connect without check valves. (*b*) Quick-connect with check valves (disconnected). (*c*) Quick-connect with one check valve (connected). (*d*) Quick-connect with check valves in both sections.

FLUID POWER COMPRESSOR AND MOTOR SYMBOLS

The compressor symbol (Figure 2.6*a*) refers to a constant displacement device. Pneumatic motors may be designed for rotation in one direction only as shown by the symbol of Figure 2.6*b*. Motors designed for rotation in either direction are identified by the symbol of Figure 2.6*c*. Figure 2.6*d* shows a symbol for a limited rotation pneumatic motor. A variable displacement pneumatic motor is indicated by the symbol of Figure 2.7*a*. A two-direction variable displacement motor is indicated by the symbol of Figure 2.7*b*. The symbol for a vacuum pump is shown in Figure 2.8.

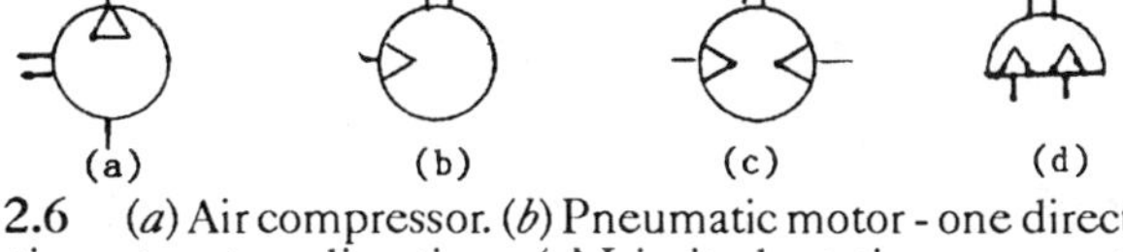

Figure 2.6 (*a*) Air compressor. (*b*) Pneumatic motor - one direction only. (*c*) Pneumatic motor - two directions. (*d*) Limited rotation pneumatic actuator.

The symbols for pneumatic compressors, motors and vacuum pumps do not indicate finite construction factors. The details of construction are relegated to maintenance manuals and company sales and service literature. History has shown that the choice of the symbol designers was correct in proposing these designs. For basic communication purposes the symbols have been kept uncomplicated. Little difficultly has been encountered with the generic characteristics of the symbol system. Several decades of use has confirmed these early decisions as being practical and efficient.

Figure 2.7 (a) Pneumatic motor - variable displacement one direction. (*b*) Pneumatic motor - variable displacement two direction.

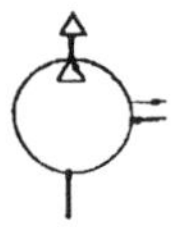

Figure 2.8 Vacuum pump.

HYDRAULIC CYLINDERS (LINEAR MOTORS)

Linear motors, consisting of a piston and cylindrical housing, are represented by the symbols shown in Figure 2.9. *Diaphragms* and other linear *actuators* are represented by similar symbols. A *single acting cylinder* is shown in Figure 2.9*a*. Air enters the left end of the cylinder and forces the piston to the right. An external force, such as gravity acting on moving machine members, produces reverse movement of the piston, forcing the air back into the line and through the control valve to the atmosphere. Figure 2.9*d* shows a modification of the single-acting cylinder fitted with an *internal spring* to return the piston to its original position. The spring space or *cavity* has an external port to the atmosphere, sometimes with a filter to prevent entrance of dirt or other contaminants.

A *double-acting cylinder* with a single rod end is shown in Figure 2.9*b*. Sometimes it is desirable to use both ends of the piston rod. To achieve this, the piston rod is extended to pass through both ends of the cylinder (Figure 2.9*c*). This arrangement is termed a *double-acting, double-rod cylinder*.

Integral devices to decelerate the piston and its associated load at the ends of the stroke are illustrated by the rectangle shown attached to the piston in Figure 2.9*e*. The addition of the slash arrow (Figure 2.9*f*) shows that the cushion or deceleration device is adjustable.

If the relationship of rod diameter to cylinder bore is significant, it may be desirable to indicate this, as in Figure 2.9*g*.

Telescopic cylinders, as the name suggests, provide long strokes by the use of multiple sections which are nested within the concentric assembly as they retract. Figure 2.9*h* shows a single-acting assembly. Figure 2.9*i* shows a double-acting telescopic cylinder. Combinations of two basic cylinders with different-size pistons are used to provide hydraulic pressure intensification or inversely, volume amplification. Air may be applied to the large end to pressurize the oil at the other end. A single, common rod may be employed. The symbols shown in Figure 2.9*j* through *l* are not intended to show construction. They indicate a pressure-intensification function or, under inverse conditions, a volume amplification.

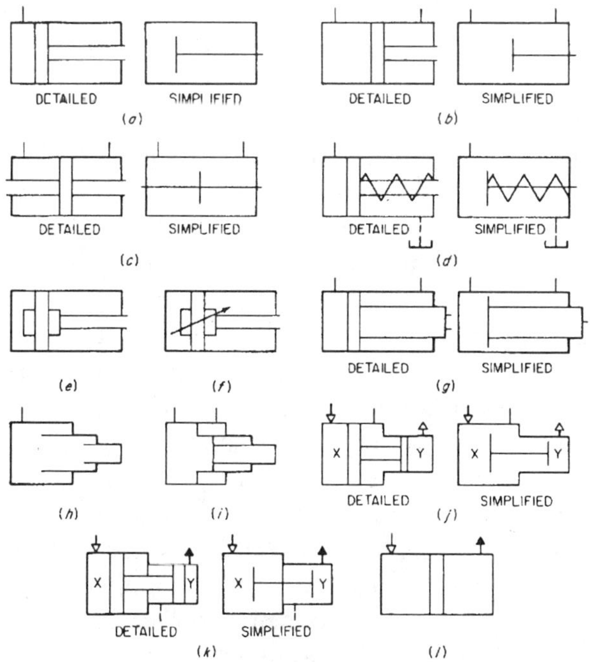

Figure 2.9 (*a*) Single-acting cylinder, diaphragm, or other linear actuator. (*b*) Double-acting, single-rod cylinder. (*c*) Double-acting, double-rod-end cylinder. (*d*) Double-acting cylinder, pressure in one direction, spring return, single-rod cylinder with drain above liquid level. (*e*) Cylinder assembly with fixed cushion, advance and retract. (*f*) Adjustable cushion, advance and retract. (*g*) Oversize rod. (*h*) Telescoping cylinder, single-acting. (*i*) Telescoping cylinder, double-acting. (*j*) Pressure intensifier or volume amplifier, pneumatic. (*k*) Hydraulic and pneumatic intensifier. (*l*) Air-oil transformer.

The assembly in Figure 2.9*j* shows that a pneumatic pressure *x* is transformed into a higher pneumatic pressure *y*. Figure 2.9*k* shows that a pneumatic pressure *x* is transferred into a higher *hydraulic* pressure *y*.

Figure 2.9*l* shows an air-oil actuator. Equipment of this design transforms a pneumatic pressure into a substantially equal hydraulic pressure, or vice versa. Such equipment is often used to obtain a smooth flow by the addition of a hydraulic-flow control mechanism. Typical units are available for power feeds on drill presses, etc.

SYMBOLS FOR DIRECTIONAL VALVES

The basic symbol for a directional valve is a rectangle, termed the *valve envelope*. As with pumps and motors, the valve envelope represents the valve enclosure or body. Lines within the envelope show the flow directions between the valve inlet and outlet openings, termed *ports*. To show the change in flow conditions when the valve is opened or closed (*actuated*), two systems of symbols are used: (1) single envelope and (2) multiple envelopes.

Single-envelope symbols are used where only one flow path exists through the valve. Flow lines within the envelope indicate static conditions when the actuating signal is not applied. The arrow can be visualized as moving to show how the flow conditions and pressure are controlled as the valve is actuated.

Multiple-envelope symbols are used when more than one flow path exists through the valve. The envelopes can be visualized as being moved to show how flow paths change when the *valving element* in an envelope is shifted to its various positions.

Figure 2.10*a* shows a typical single envelope; multiple envelopes are shown in Figure 2.10*b*. Flow lines to the valve ports are added to the two types of basic valve envelopes in Figure 2.10*c* and 2.10*d*. Note in Figure 2.10*d* that the ports are shown at the center envelope segment. The flow conditions are shown connected to the rest position envelope if only two envelopes are used omitting the center envelope.

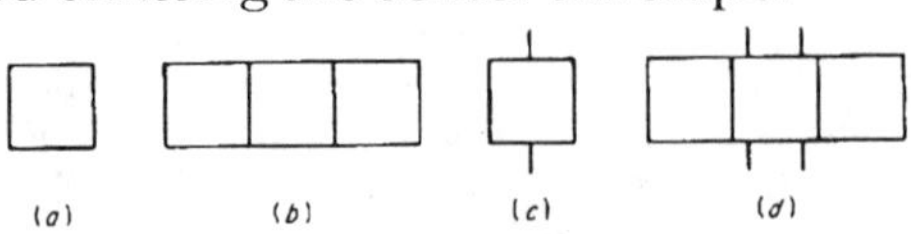

Figure 2.10 (*a*) Basic envelope. (*b*) Multiple basic envelopes. (*c*) Ports attached to basic envelope. (*d*) Ports attached to normal multiple basic envelopes.

A variety of valve-port conditions are shown in Figure 2.11. Thus, the double arrow in Figure 2.11*a* denotes a blocked or closed port in a single envelope where flow may be expected to pass in either direction when the valve is actuated. The blocked condition is indicated by the fact that the arrows do not line up with the lines coming to the single square. Figure 2.11*b* shows the same condition in a multiple-envelope valve. The single envelope signifies that an infinite number of positions may be encountered between the closed (normal) and open position, such as might be expected with a relief valve where the opening is a function of the orifice size that must be created to provide the needed pressure level in an automatic mode.

The multiple envelope of Figure 2.11*b* and *c* indicates a finite number of positions which the directing elements can be normally expected to assume. The normal position of Figure 2.11*b* is usually indicated by the left block. The normal position of the unit, shown in Figure 2.11*c*, is the neutral position in which all the ports are blocked. This condition is referred to as a *closed center*. Shifted to conditions in the left block, the flow is in the direction indicated by the arrows. When shifted to the right, the flows are reversed. This is referred to as a *four-way function*. One input port can be directed to separate outlets or blocked, while a return line is directed to an opposite port in a predetermined pattern.

A through flow is indicated by the symbol of Figure 2.11*d*. An infinite number of positions of the flow-control elements from full-open to closed can be anticipated with this design. The two-element symbol of Figure 2.11*e* indicates two finite positions. The three-element symbol of Figure 2.11*f* indicates three finite positions, with the neutral (rest) position indicating the full interconnection of all associated lines. Pneumatic valves usually employ two positions for a three or four-way valve.

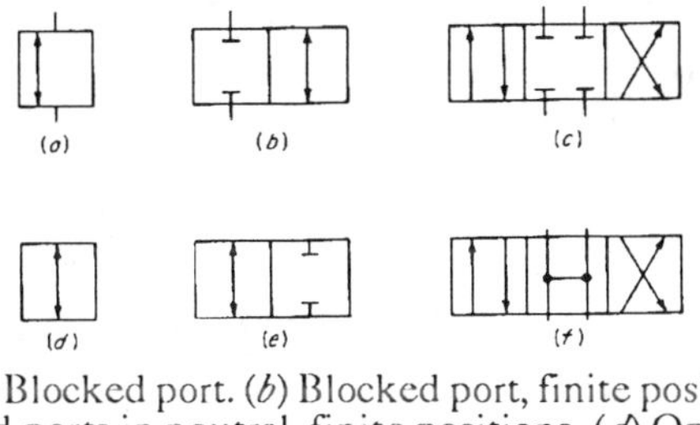

Figure 2.11 (*a*) Blocked port. (*b*) Blocked port, finite positions. (*c*) Three-position, blocked ports in neutral, finite positions. (*d*) Open port, infinite positions. (*e*) Normally-open, finite positions. (*f*) Open center, three finite positions.

Figure 2.12*a* and *b* summarize the basic symbol, normal and actuated conditions for normally-open and normally-closed single envelope valves. Figure 2.12*c* shows an alternate arrow design which indicates that one port is always connected to one flow path, whether normal or actuated.

The symbol for a reducing valve or *pressure regulator* shows the supply line connected to the downstream line in the normal rest position as in Figure 2.12*a*. The pilot connection is downstream as shown in Figure 2.13*a*.

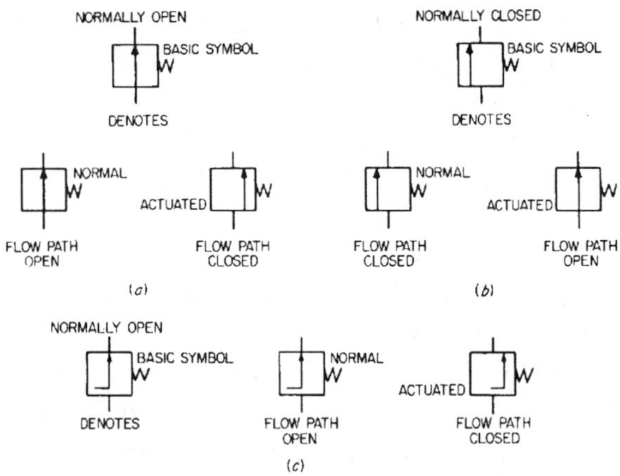

Figure 2.12 (*a*) Normally-open, single-envelope valve. (*b*) Normally-closed, single-envelope valve. (*c*) Alternate arrow design to indicate that one port is always connected, whether actuated or at rest.

The valve in Figure 2.12*b* could be a *spring-loaded relief valve*, piped so that the discharge flows to the atmosphere. The signal source for a relief valve is ahead of the normally-closed element. As the pressure rises to a value at which the valve opens and the flow passes through the valve, it can be imagined that the arrow is aligned as in Figure 2.12*b*, where it is shown in the actuated position.

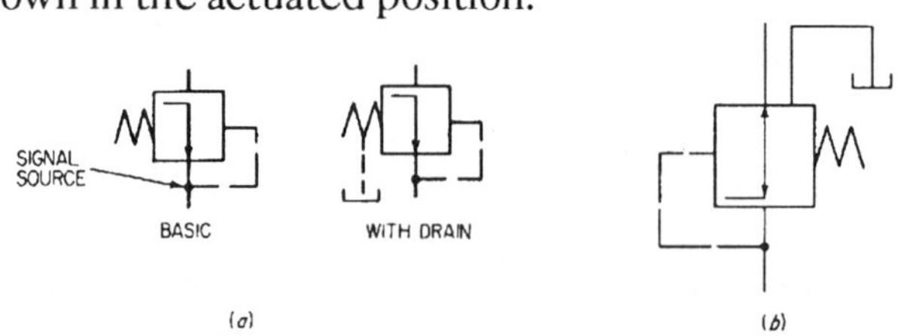

Figure 2.13 (*a*) Pressure regulator with signal from downstream. (*b*) Pressure regulator with integral relief function.

Figure 2.13*b* illustrates a pressure regulator with a relief valve function. Any pressure above the adjusted value is relieved to the atmosphere. Figure 2.14 illustrates a check valve which allows free flow in one direction and stops flow in the other direction.

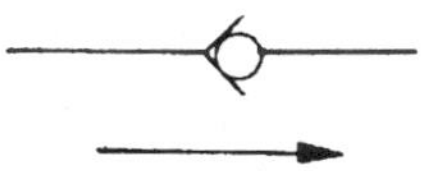

Figure 2.14 Check valve.

CONDITIONING OF THE AIR

The symbol for a filter or strainer employs a dashed line bisecting the square (Figure 2.15*a*). No attempt is made to differentiate between the various types of filters or strainers.

The position of the symbol in the circuit will usually provide a clue as to the type. Filters may be found in any part of a circuit. There are no fixed rules as to strainers and filter usage, so the material list on a drawing may be needed to indicate the specific information. Experience has proven that a clean system, well conditioned air, and properly positioned and serviced filters lead to long, trouble free service from the pneumatic power transmission system.

Much pneumatic equipment is used in association with hydraulic systems. Because of this, it is usual to find pneumatic conditioning device symbols on hydraulic drawings. Pneumatic directional-control valves, used in circuitry, employ symbols similar to the equivalent hydraulic function. Hollow, equilateral, energy-flow triangles provide the clue as to the fluid medium. Figure 2.15*b* shows a separator used to remove moisture from pneumatic lines. The symbol of Figure 2.15*c* illustrates a combination filter and separator for pneumatic lines. A desiccator, or chemical drier, is shown in Figure 2.15*d*.

Pneumatic lines usually require the introduction of a lubricant into the line to minimize rust and corrosion. The oil may also serve to assist in lubricating operating devices in the circuit, such as air-piloted hydraulic valves. Figure 2.15*e* shows the units with and without a manual drain.

Figure 2.15*f* shows the symbol for an electric-drive motor. The symbol for Figure 2.15*g* illustrates a heat engine, such as a steam, diesel or gasoline engine.

If a circuit drawing is provided with the required flow and pressure data it is generally possible to identify components which may be the

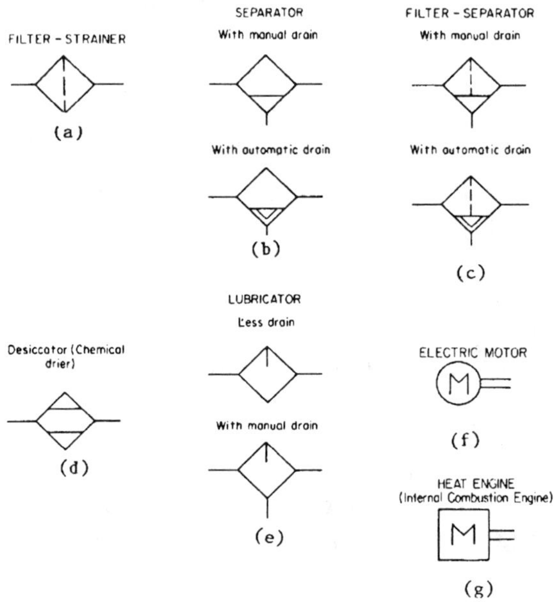

Figure 2.15 (*a*) Filter strainer. (*b*) Separator. (*c*) Filter separator. (*d*) Desiccator. (*e*) Air line lubricator. (*f*) Electric motor. (*g*) Heat Engine.

cause of a malfunction. Information must also be provided relative to the available flow.

After identifying a component which could be the cause of a malfunction and making *sure* there is no danger of a sudden motion of any actuator, it may be possible to insert or connect a pressure gage to make certain pressurized air is available to the valve. Pressure beyond the valve at the actuator may indicate a jammed machine member needing mechanical attention. Because of its compressibility – air under pressure can be misleading – an unwanted restriction limiting the flow of air in the line can be the problem.

Significant changes in local operating pressure can indicate a friction build-up requiring better lubrication, proper realignment or possibly a major repair. A basic knowledge of electrical circuitry is helpful to insure that a suitable electric signal is being provided at the proper time to achieve the desired machine circuit sequence. Obviously, the maintenance personnel must be acquainted with the machine operating manual in addition to the needed lubrication, filtration and other basic adjustments.

The operating manual for a new machine should have provisions to accept data developed when the machine is set-up to run in the desired manner. If the machine manufacturer has not provided this information - get it either from the manufacturer or make a suitable circuit drawing - indicating pressure and flow information and keep it for reference. This should include pressure readings, etc. which should be noted on the circuit drawing. In some equipment, a flow rate indicating device may be installed to monitor current performance and to provide data on "normal" pressure and flow.

Water in a system is not compressible. It can change the operating values. Contamination in a pneumatic system can be water, oily residue or solids - such as sand. Water in a liquid form entering a cushioned cylinder from any source can be disastrous - *and dangerous* - especially in high speed circuits. Just imagine the air cylinder cushions set for air suddenly faced with water to stop on! Or - flow across the needle valve of a "flow control valve" suddenly being a slug of water (Figure 2.16). Results - a machine out of adjustment and sometimes damaged beyond repair.

Figure 2.16 A slug of water in a restricted area can act like a formidable barrier. (Courtesy of Ross Operating Valve Company)

The operating manual with the initial pressure and flow values must be readily available to maintenance and service personnel. A record of maintenance and repair functions with operating pressure and velocity data noted on the circuit drawing showing normal operating data while the machine is in service and operating well can be very useful to maintenance and repair personnel when tracing the cause of malfunctions.

SUMMARY

Operator training must include basic safety considerations. In addition operator/maintenance personnel must be acquainted with the circuit drawings, operating characteristics and normal operating pressures and speed rate.

Typical normal operating pressures added to the circuit drawing aids in assessing malfunctions. Operators should report unusual operating sound patterns or erratic movements to the maintenance personnel for corrective actions.

REVIEW QUESTIONS

1. What information is provided by pneumatic symbols?
2. What international organization is responsible for pneumatic symbol technology?
3. How can you determine whether an interconnecting line or component symbol is for a hydraulic or pneumatic system?
4. What type of device is used to convert pneumatic power to hydraulic power?
5. What is the purpose of introducing lubricants to air lines?
6. Is a pressure regulator considered a normally open valve?
7. Why are multiple-envelope valve symbols used?
8. Describe a telescopic cylinder.
9. What is added to a motor symbol to indicate variability?
10. What organization monitors the development and use of pneumatic symbols in the United States?

3

THE GAS LAWS

THE AIR AROUND US

With each average breath, you are inhaling a little over 3 percent of a cubic foot of air - a colorless, odorless, tasteless mixture of gases without which you could not survive for more than five or six minutes. This air that we breathe is the air we compress and use in an endless variety of ways vital to our industrial community. Present knowledge of air is the result of centuries of experience and experiments, inspired guesswork and hard labor by many scientists and philosophers.

The Greeks, in Aristotle's time considered air to be one of the basic elements out of which the universe was structured. In Aristotle's view there was first of all earth. Where earth ended, water began. Where the water ended, there was air. Above the air was the element of fire. Fire was manifested in lightning and the occasional meteors. Where fire ended, it was assumed a perfect kind of substance began which permeated the rest of the universe. The Greeks named this "aither", a term to indicate incorruptible from which we obtained our word ethereal. This concept utterly dominated this part of science, as well as most other scientific thought. For almost nineteen centuries nothingness had no place, and the expression "nature abhors a vacuum" still persists, rooted in this early view of the world and the universe.

When the first hole was excavated in a mining activity there was the common problem of water seepage into the bottom of shafts and excavations. Simple hand pumps were devised to bring this water to the surface. Virtually all pumps consisted of a cylinder of sorts into which a piston was fitted. A crude conduit, usually fabricated of wooden pipe sometimes covered with animal entrails, was connected between the bottom of the cylinder and the pool of water. A handle attached to the piston provided the motive power. When the handle was pushed down, the piston could be raised in the cylinder, creating a vacuum. Under the cylinder this pipe of sorts was extended down below the surface of the water. By means of crude but effective flapper check valves the water entered the cylinder to fill the vacuum and was subsequently pushed out by the piston on the next stroke of the handle. Not too different from the hand pumps encountered in roadside parks that delight modern day youngsters who look forward to a stream of cool, clear water after vigorously moving the handle up and down.

It was soon noted that no matter how vigorously the handle was pumped or how well the piston was fitted to the cylinder, the water could not be raised higher than about thirty feet. This failure of suction was of interest to one of Galileo's students by the name of Evangelista Torricelli. Evangelista suspected that the suction of the pump was not responsible for pulling the water up, but rather that the water was pushed into the cylinder by the pressure or weight of the air once a partial vacuum was established by the pump. He suspected that the air had a finite weight equal to a specific column of water resulting in this limitation of the pumping action.

As a test Torricelli filled a glass tube about a yard long with mercury, sealing one end of the tube. Holding his thumb over the open end of the tube, he upended the tube into a dish partially filled with mercury. On removing his thumb, it must have been an extraordinary moment for Torricelli to watch the mercury flow out of the tube into the dish and then stop flowing when the column fell to about thirty inches high. The weight of the air (atmosphere) exerted a force equal to thirty inches of mercury (Figure 3.1).

THE FIRST BAROMETER

Since mercury is 13.6 times as dense as water, this equals about 34 feet of water. If a good enough vacuum could be established, 33.9 feet is the highest the atmosphere would lift a column of water under these circumstances. This helped to explain the mine-and-pump problem. The experiment resulted in the invention of the barometer and also created the first decent man-made vacuum. The space left above the mercury at the top of the tube is still known as a "Torricellian" vacuum.

The question then arose as to whether the air had the same weight all the way to the top, where the "fire" began. If so, it can be calculated that the height of the atmosphere would have to be about five miles. Since there seemed no way to verify this, short of visiting these high altitudes, this view was generally accepted. Robert Boyle, an English physicist in the latter part of the 17th century, in another experiment using glass tubes and mercury established the relationship between the pressure and the volume of air.

He partially filled a "J"-shaped glass tube with mercury. He then plugged the short leg and proceeded to pour additional mercury into the long leg of the tube. He observed that when he doubled the exerted pressure of the mercury in the long leg of the "J", the air space above the mercury in the short leg was reduced by half (Figure 3.2). This experiment indicated that as he doubled the pressure, he reduced the volume of the air by half. This is Boyle's Law, one of the two principal laws comprising the ideal or general gas law. Boyle's Law states that "with a constant temperature, volume is inversely proportional to pressure".

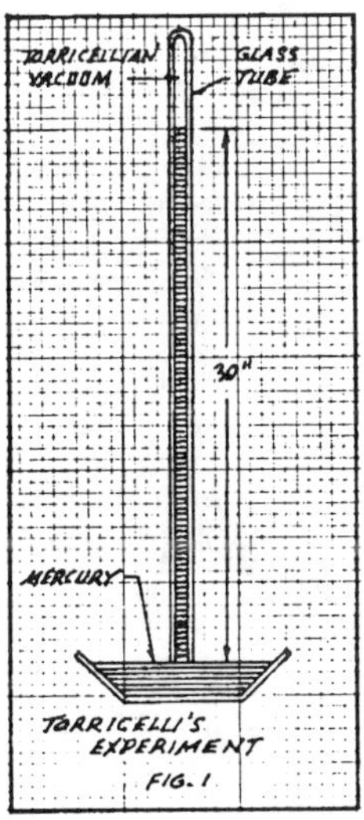

Figure 3.1 Torricelli's Experiment. (Courtesy of Wilkerson Corporation)

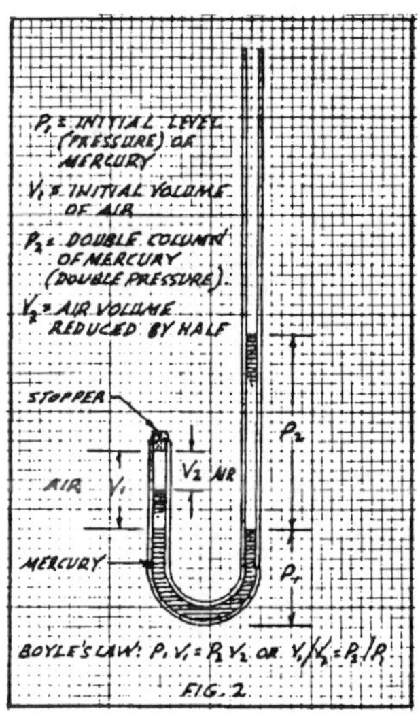

Figure 3.2 Boyle's Law. (Courtesy of Wilkerson Corporation)

SEA LEVEL WAS A GOOD STARTING POINT

By proving that air contracts under pressure, it would follow that it must be most dense at sea level and steadily gets thinner as the overlaying air decreases. Blaise Pascal, the French physicist at about the same time empirically demonstrated this by comparing the column of mercury in a "Torricellian Barometer" at the base of a mountain in central France and again at about a 3,000-foot elevation on the slopes of the mountain. He noted a three-inch difference in the column. It has been shown that the weight of the atmosphere decreases 10-fold with every 12 miles of altitude. At sea level the weight of the air is equal to 30 inches of mercury. At 12 miles it would be equal to 3- inches of mercury; 24 miles high, 0.3-inches: and so forth. Calculating this on out at 108 miles above the earth, the weight of the remaining atmosphere is a minuscule 0.00000003-inch of mercury. Nevertheless, there still remains over 6 million tons of air above 108 miles.

THE GASES THAT MAKE UP OUR ATMOSPHERE

Experiments conducted through the 18th century by various men such as Priestly, Rutherford, Cavendish and Lavoisier established that air was a mixture of at least four chemically different gases. Names were coined and given to these various gases. Henry Cavendish, for example, in an effort to isolate and prove the existence of an imponderable substance called "phlogiston", discovered hydrogen as one of the trace elements in the air. Phlogiston was a fluid material invented by a German physician and chemist, George Stahl, around 1700 to help explain the behavior of heat transfer, combustion, and temperature phenomena.

When the French physicist J. A. C. Charles appeared, it was known that air was a combination of nitrogen and oxygen in a proportion of approximately four-fifths to one-fifth, along with traces of other gases. Jacques Alexandre Cesar Charles lived in the late 1700's when the investigation of heat, combustion and temperature was of comparable import and interest as was electricity in the 1800's and radioactivity in the early 1900's. Charles's most memorable discovery was in establishing the relationship of temperature to the pressure and volume of air and other gases. He found that starting at zero degrees Celsius, for every one degree increase (or decrease) in temperature, the volume of the air increased or decreased by 1/273. While there seemed no problem of logic in the temperature increase aspect, cooling the air presented an apparent paradox. Namely: starting at zero degrees Celsius, when one got to -273 degrees Celsius, the original volume of air would disappear. Scientists at the time were sure this wouldn't happen, but they had no explanation why. With the development of atomic theory this paradox was viewed in new terms. Namely that temperature was a measure of average molecular velocity, and that at some point the molecules would not have enough kinetic energy to remain as a gas; would turn to liquid, and liquid behaves altogether differently than a gas, etc.

From his investigation and experiments, Charles gave his name to the other principal law - along with Boyle - which makes up the ideal gas law. Charles's Law was officially enunciated in 1787, and one way of stating this law is "with a constant volume, absolute pressure is directly proportional to absolute temperature." That is to say, if you double the absolute temperature, keeping a constant volume, you will double the pressure (Figure 3.3). Charles's Law is in mind when you are cautioned against throwing empty aerosol cans into a fire. As the remaining gas in the can is heated, the pressure might increase sufficiently to explode the container.

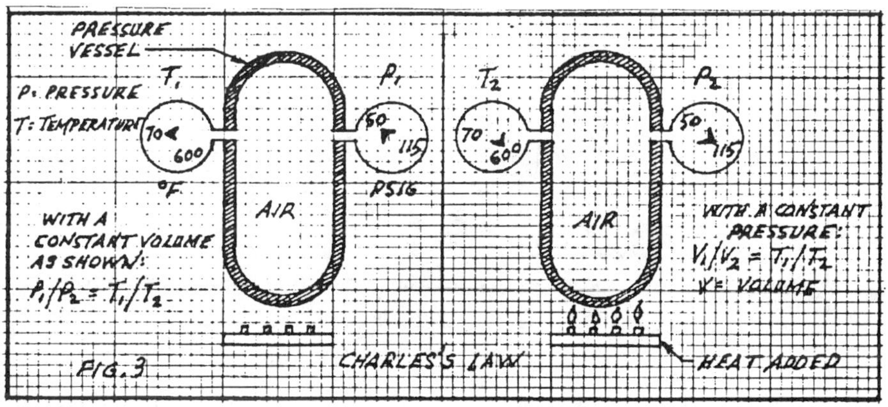

Figure 3.3 Charles's Law. (Courtesy of Wilkerson Corporation)

With the invention of spectroscopy in the closing years of the 19th century, William Ramsey isolated and named most of the trace gases in the air, such as Argon (inert), Neon (new), Krypton (hidden), Xenon (stranger), and so forth. In recent decades, nitrous oxide (laughing gas), carbon monoxide, and methane have been discovered in the atmosphere.

DRY AIR AT SEA LEVEL

Figure 3.4 shows the composition of dry air at sea level under normal conditions. Small amounts of extraneous impurities present in all atmospheric air may differ from place to place - this might be especially noticeable around heavy industrial areas or frequently traveled highways. Water vapor or water "gas" is not shown in Figure 3.4, because it varies so much from one location to another and from time to time. Water vapor is an important part of the air, affecting its density and its dynamic behavior to a very great extent. As a matter of fact, a critical part of our specification of "standard" air is the water vapor content.

GAS	SYMBOL	% BY VOLUME
NITROGEN	N_2	78.048
OXYGEN	O_2	20.9476
ARGON	A	0.934
CARBON DIOXIDE	CO_2	0.0314*
NEON	Ne	0.001816
HELIUM	He	0.000524
METHANE	CH_4	0.0002*
KRYPTON	Kr	0.000114
HYDROGEN	H_2	0.00005
NITROUS OXIDE	N_2O	0.00005
XENON	Xe	0.0000087

COMPOSITION OF DRY AIR AT SEA LEVEL

* VARIABLE

Figure 3.4 Composition of dry air at sea level. (Courtesy of Wilkerson Corporation)

Since the earliest days of chemistry, chemists were intrigued by the three states of any substance: solid, liquid and gaseous - the most common example being water. As a solid, it is ice; in its liquid state, it fills the rivers, lakes, and oceans; and when it evaporates, it becomes a gas. In each state, it is chemically the same. As a gas, water (or any substance) is broken down so that the individual molecules are comparatively free of each other. The water molecules consist of two hydrogen atoms and one oxygen atom. Any further breakdown would represent a chemical change, and you would no longer have water.

Why is it we refer to water in its gaseous state as "water vapor" instead of "water gas"? In the early 1600's, Jean Baptista van Helmont, a Flemish chemist (who is credited with coining the word "gas") noted that some substances remained a gas through a wide temperature range, such as carbon dioxide - which, in fact, refused to be other than gas. However, liquid water, when heated, seemed to turn into steam before it disappeared and could easily be reconverted into liquid by subsequent cooling - all through a small temperature range. True enough, left in the open, liquid water would disappear (evaporate) without any visible steam. Nevertheless, van Helmont referred to water gas as "vapor", the word vapor stemming from a Greek word meaning "steam". This differentiation stuck, and we still often refer to water vapor rather than water gas. (A more refined differentiation now exists but is not cogent to this text). It is well to remember, however, that the water vapor in the air is as gaseous as the oxygen or nitrogen gas, for example. (Visible steam, incidentally, is not water vapor. The steam that we can see consists of individual particles of liquid water).

The amount of water vapor in the air is described by the words "relative humidity". The capacity for air to hold water in vapor or gaseous form is primarily a function of the temperature of the air, so any explanation of relative humidity must make specific reference to the temperature of the air. The higher the temperature, the greater the capacity to hold water vapor. Figure 3.5 is a chart showing water vapor content of air at various temperatures and percentages of saturation. Briefly, relative humidity means the amount of water *vapor* in the air at a given temperature, relative to its total capacity at the same temperature. Note from Figure 3.5 that as the temperature of the air drops the capacity to hold water vapor drops. By withdrawing heat energy from air you can dry air - that is, condense a portion of the water vapor. In fact, heat exchangers are used extensively as devices for drying air.

Using Figure 3.5 and starting with air at 70 degrees Fahrenheit and 60% relative humidity, note there are 4.79 grains of water vapor per cubic foot. The air still has a capacity to hold *more* water vapor, because it is only 60% saturated. How much more water vapor? 3.19 grains more per cubic foot. This is the difference between the 4.79 grains at 60% and the 7.98 grains which it can hold at saturation or 100% relative humidity. 100% relative humidity is also the dew point of the air. This is the temperature point where any water vapor will begin to transform, or

WILKERSON CORPORATION

WEIGHT OF WATER IN A CUBIC FOOT OF AIR AT VARIOUS TEMPERATURES AND PERCENTAGES OF SATURATION

7,000 grains = 1 lb. Weights shown in grains.

TEMP. °F.	RELATIVE HUMIDITY 10%	20%	30%	40%	50%	60%	70%	80%	90%	100%	TEMP. °F.
-10	.028	.057	.086	.114	.142	.171	.200	.228	.256	.285	-10
0	.048	.096	.144	.192	.240	.289	.337	.385	.433	.481	0
10	.078	.155	.233	.310	.388	.466	.543	.621	.698	.776	10
20	.124	.247	.370	.494	.618	.741	.864	.988	1.11	1.24	20
32	.211	.422	.634	.845	1.06	1.27	1.48	1.69	1.90	2.11	32
35	.237	.473	.710	.946	1.18	1.42	1.66	1.89	2.13	2.37	35
40	.285	.570	.855	1.14	1.42	1.71	1.99	2.28	2.56	2.85	40
45	.341	.683	1.02	1.37	1.71	2.05	2.39	2.73	3.07	3.41	45
50	.408	.815	1.22	1.63	2.04	2.45	2.85	3.26	3.67	4.08	50
55	.485	.970	1.46	1.94	2.42	2.91	3.39	3.88	4.36	4.85	55
60	.574	1.15	1.72	2.30	2.87	3.45	4.02	4.60	5.17	5.75	60
65	.678	1.36	2.03	2.71	3.39	4.07	4.75	5.42	6.10	6.78	65
70	.798	1.60	2.39	3.19	3.99	4.79	5.59	6.38	7.18	7.98	70
75	.936	1.87	2.81	3.74	4.68	5.62	6.55	7.49	8.42	9.36	75
80	1.09	2.19	3.28	4.37	5.47	6.56	7.65	8.75	9.84	10.93	80
85	1.27	2.54	3.81	5.08	6.35	7.62	8.89	10.16	11.43	12.73	85
90	1.48	2.96	4.44	5.92	7.40	8.87	10.35	11.83	13.31	14.78	90
95	1.72	3.44	5.16	6.88	8.60	10.32	12.04	13.76	15.48	17.15	95
100	1.98	3.95	5.93	7.91	9.88	11.86	13.84	15.81	17.79	19.77	100
110	2.63	5.26	7.89	10.52	13.15	15.78	18.41	21.04	23.67	26.33	110
120	3.45	6.90	10.35	13.80	17.25	20.70	24.15	27.60	31.05	34.48	120
130	4.44	8.88	13.32	17.76	22.20	26.64	31.08	35.52	39.96	44.42	130
140	5.68	11.36	17.04	22.72	28.40	34.08	39.76	45.44	51.12	56.82	140
150	7.19	14.38	21.57	28.76	35.95	43.14	50.33	57.52	64.71	71.96	150
160	9.04	18.08	27.12	36.16	45.20	54.24	63.28	72.32	81.36	90.44	160
170	11.26	22.52	33.78	45.04	56.30	67.56	78.82	90.08	101.34	112.63	170
180	13.92	27.84	41.76	55.68	69.60	83.52	97.44	111.36	125.28	139.23	180

Figure 3.5 Water content of air at various temperatures and percentages of saturation. (Courtesy of Wilkerson Corporation)

condense, into liquid "dew". Notice from the chart (using the previous example) that the dew point of air (at 70 degrees Fahrenheit and 60% RH) is about 54 degrees Fahrenheit. You find this by noting where in the vertical 100% relative humidity column the water vapor content of the air at 70 degrees Fahrenheit and 60% RH (which is 4.79 grains) represents saturation. In other words, 4.79 grains of water vapor is *all* the air can hold (saturated - 100% relative humidity) at approximately 54 degrees Fahrenheit.

STANDARD AIR AND FREE AIR

In the United States, air is most commonly measured by the cubic foot. As discussed earlier, the dynamics of air is for the most part determined by the air's pressure, temperature, and water vapor content (relative humidity). Therefore, American industry has adopted a standard on air specifying these three factors. Pressure differentials, pressure drop, flow, and performance criteria of various pneumatic equipment and components are based and designed around this "standard" air. A standard cubic foot (scf) of air is: one cubic foot volume at sea-level atmospheric *pressure* (14.69 pounds per square inch), 68 degrees Fahrenheit temperature, and 36% relative humidity. This is the standard established by ASME (American Society of Mechanical Engineers). For air to perform some useful work, there must be pressure differentials. Flowing air (a function of pressure difference) is measured by the cubic foot volume past a given point during some period of time, usually by the minute; so we have scfm or standard cubic feet per minute. You may occasionally see the initials scfh (standard cubic feet per hour) the same as scfm but with a longer time factor. The amount of air flowing past some point is occasionally called out by weight. For example, a flow of 20 lbs of standard air per minute. Standard air weighs 0.0762 pounds per cubic foot, so 20 pounds converted to cubic feet is 261.6 cubic feet. (If you worked this out, you would find there is 13.08 scf per pound). Contrary to what most people think, the drier and cooler air is, the *more* it weighs. Hot, humid air weighs significantly less than cold, dry air because of the activity of the heat within the molecular structure. Molecular activity is reduced and greater mass accrues as the temperature decreases. Aircraft pilots must be constantly aware of this when calculating required take-off distance, aircraft loading, etc. It is well to make a distinction between the *weight* of air and the *pressure of air*. Weight is the air's density or mass. A truck wheel and tire that weigh 150 pounds empty, assuming it will take 13.08 cubic feet of air to inflate the tire, will weigh 151 pounds when inflated. Pressure is the force that air exerts against a surface, which in the United States is by the square inch: pounds per square inch (psi). Even though these two terms are interrelated - the weight of the atmosphere pressing down at sea level exerts a force equal to a column of mercury slightly less than 30 inches high, equivalent to 14.69 psi. These terms should not be confused.

While it is well to know the definition of a standard cubic foot of air, you will rarely encounter air under these precise conditions. Rather, you will be normally dealing with "free" air. Free or "actual" air is ambient air with its constantly changing temperature and relative humidity. ("Ambient" means surrounding or encompassing.) Atmospheric pressure does, of course, change with altitude, but in any one area it remains essentially constant. For most practical infield applications of compressed air and compressed air equipment, "free" air is equal to "standard" air.

Starting with most ambient air conditions around a compressor, free air - when compressed to normal in plant pressures - will result in saturated air at the storage tank. So whether you start with standard air or free air, the practical application of the components we're concerned with will essentially be the same. Granted, changes in elevation, relative humidity, and temperature can and do affect compressor performance. For example; once the air is compressed and ready for use in the storage tank, its original condition is practically academic.

To understand this we must consider a few of the things that happen to air when it is compressed. Air is "springy", and this compressibility of air means you can keep squeezing more and more air into a constant volume (or storage tank). As you force more air into this constant volume, the pressure rises - this is Boyle's Law. This ratio of free air required to achieve elevated pressures in a constant volume is expressed by the following equation:

$$\frac{\text{ATMOSPHERIC PRESSURE + GAUGE PRESSURE}}{\text{ATMOSPHERIC PRESSURE}} \quad \text{or} \quad \frac{\text{PSIA}}{\text{ATMOS.}}$$

Therefore, if we want to know how many cubic feet of free air we must pump into a cubic foot container to get a 100 psi gauge reading, we use this equation. Pressure gauges normally show only elevated pressures. Gauge pressure is psig - pounds per square inch gauge. The sum of gauge pressure and atmospheric pressure is called absolute pressure: abbreviated; psia - pounds per square inch absolute. Knowing and using the absolute pressure is also required in calculating flow and pressure drop on various components. At sea level, this ratio would be:

$$\frac{\text{100 psig} \;=\; \text{14.7 atmos}}{\text{14.7 atmos}} = \frac{\text{114.7 psia}}{\text{14.7 atmos}} = \text{7.8 (A ratio of 1 to 7.8)}$$

and at a 5,000-foot elevation, we would have:

112.22 + 9.18, a ratio of 1 to 9.18

See Figure 3.6 for atmospheric pressure at various altitudes.

ALTITUDE ABOVE SEA LEVEL FEET	PSIA	INCHES Hg
0	14.69	29.92
500	14.42	29.38
1000	14.16	28.86
1500	13.91	28.33
2000	13.66	27.82
3000	13.16	26.81
4000	12.68	25.84
5000	12.22	24.89
6000	11.77	23.98
8000	10.91	22.22
9000	10.50	21.38
10,000	10.10	20.58

Figure 3.6 Atmospheric pressure at various altitudes. (Courtesy of Wilkerson Corporation)

These equations tell us that at sea level, it takes about 8 (let's round the 7.8 to 8) cubic feet of free air to make one cubic foot at 100 psig.

Let's take an example using ambient air at sea level pressure (14.69 psia) 70 degrees Fahrenheit temperature, and 50% relative humidity. Under these conditions, Figure 3.5 shows that each cubic foot will contain 3.99 grains of water vapor. From this equation, we know we will have to pump 7 cubic feet of free air into a 1 cubic foot storage tank to get 100 psig in the tank (remember, there was already 1 cubic foot of free air in the tank to start with). Figure 3.7 is a chart showing the ratio of compression and is equal to the equation shown above. There will be heat generated as we force this air into our storage tank. However, through heat exchangers of one sort or another, this heat of compression is mostly dissipated, and we generally find that normal temperatures of the air at the storage tank will be from 100 degrees Fahrenheit to 120 degrees

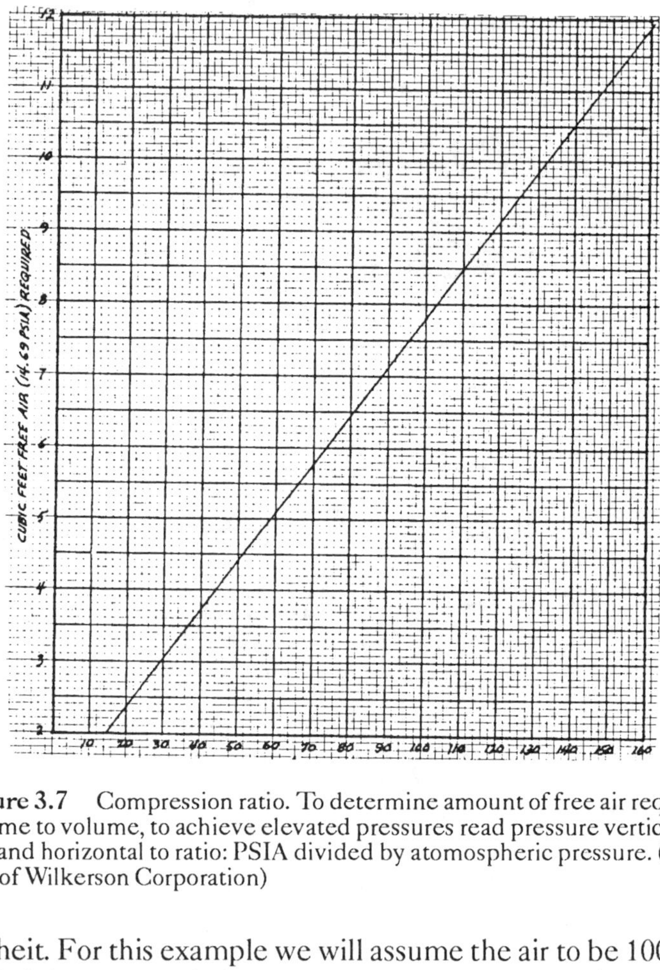

Figure 3.7 Compression ratio. To determine amount of free air required, volume to volume, to achieve elevated pressures read pressure vertically to line and horizontal to ratio: PSIA divided by atomospheric pressure. (Courtesy of Wilkerson Corporation)

Fahrenheit. For this example we will assume the air to be 100 degrees Fahrenheit in our 1 cubic foot storage tank.

At 100 psig we now have 8 times as much water vapor in our tank as we had at zero gauge pressure: 8 x 3.99 grains equaling 31.92 grains total. It is a happy fact that at the pressures we will be concerned with we need *not* consider the pressure factor but only the temperature to determine how much water vapor a volume of air can hold. Whether the pressure is 0 psig or typical system pressures, the water vapor capacity of a volume of air is a function of temperature only. How much water vapor will our cubic foot storage tank hold at 100 degrees Fahrenheit? Well, it won't hold more than its saturation point or its 100% relative humidity point. Look under the 100% relative humidity column, and you will see that at 100 degrees Fahrenheit our cubic foot of air will only hold 19.77 grains, this being its *total capacity* at this temperature. Therefore, 12.15 grains of water will be converted from vapor to liquid:

31.92 grains water vapor in tank
19.77 maximum capacity water vapor @100 degrees F
12.15 grains converted to liquid

If you wish to go through this exercise using other conditions of free air, you will find that starting with almost any set of ambient conditions the compressed air in the storage tank will end up saturated at storage tank temperatures. To return to our earlier statement, then: for most practical infield applications of compressed air and compressed air equipment, "free" air is equal to "standard" air; that is, the air in the system at pressure is essentially the same from one geographical area to another and from season to season.

A REVIEW AND THE APPLICATIONS OF BOYLE'S LAW

Robert Boyle, born in Ireland in 1627, is credited with being the first person to investigate and document the exact relation between the change in the pressure exerted by a confined body of gas and its change in volume. Boyle formulated the important law now known by his name: *The absolute pressure which a given quantity of gas at constant temperature exerts against the walls of the containing vessel is inversely proportional to the volume occupied.* Absolute pressure is the gage pressure plus the atmospheric pressure (Gage + 14.7 psi in the customary unit system).

THE PNEUMATIC SPRING

Boyle's Law explains the increase in air pressure as an external force is exerted against a pneumatic cylinder with valving in a closed mode. Granted that the cylinder is closed at the end opposite the rod so the movement of the piston in the cylindrical tube is possible because the air can be compressed. The same cylinder under equal conditions with a liquid instead of the compressed air will be virtually solid with no physical movement of the piston rod because of the relative incompressibility of the liquid.

Thus a pneumatic cylinder can hold a movable roll against a fixed roll to exert a desired force against a product rolling through the machine. In the event of a hard foreign element in the product being rolled, it will be possible for the movable roll to retract momentarily to avoid damage. The roll can move as a result of the captive air in the cylinder. The air pressure will momentarily increase slightly but the roll will not be damaged. As the foreign matter passes through, the roll will return to the desired position. The sportsman's air rifle functions as a result of the captive air in the pocket behind the projectile.

Boyle's Law explains the action of the hydraulic accumulator. The chamber of the accumulator in Figure 3.8*a* includes a floating check to prevent loss of gas when the liquid is exhausted by the action of the compressed gas. This type of accumulator is used in very large presses where large volumes of liquid must be moved rapidly by the forces in the captive gas. A resilient ball separates the liquid and gas of Figure 3.8*b*. The ball-type structure is used for relatively low pressures and large flow rates. A liquid chamber enclosed in the gas chamber as shown in Figure 3.8*c* equalizes the pressure on the walls to maintain a uniform fit for the moving separator piston. This type of assembly has been used in high shock systems where the inertial values could expand the walls of a structure such as Figure 3.8*a* or Figure 3.8*b*.

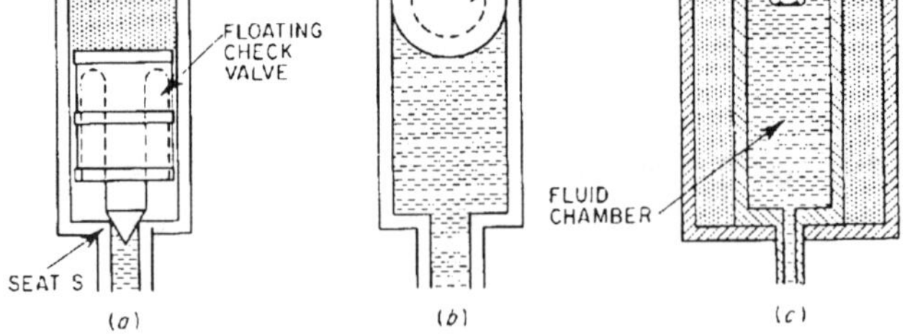

Figure 3.8 Gas may exert force directly against a liquid as in (*a*). A resilient ball may separate the gas from the liquid as in (*b*). The chamber may be structured so that the contained liquid is separated from the gas with a piston, and the fluid chamber encompassed within the gas will have minimum stresses on the wall as in (*c*).

AIRCRAFT ACCUMULATORS

The accumulators of Figure 3.9*a* were developed for aircraft use to provide the desired function in a light, compact assembly. The aircraft assembly was expanded at a later date to provide greater capacity for industrial uses by inserting a bag into an elongated shell as shown in Figure 3.9*b*.

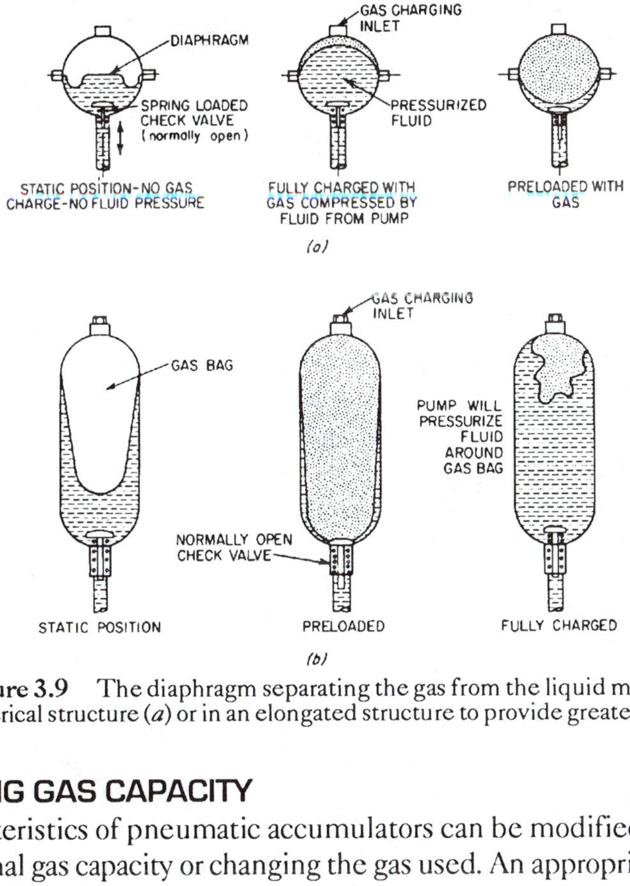

Figure 3.9 The diaphragm separating the gas from the liquid may be in a spherical structure (*a*) or in an elongated structure to provide greater capacity (*b*).

ADDING GAS CAPACITY

Characteristics of pneumatic accumulators can be modified by adding additional gas capacity or changing the gas used. An appropriate gas tank of needed capacity can be connected to the gas side of the spherical accumulator of Figure 3.9*a* or to the bag of the accumulator shown in Figure 3.9*b*. This can be very important if a narrow working pressure range is desired.

TEMPERATURE CONSIDERATIONS

French chemist Jacques Charles in 1787 experimented with a constant volume of gas. He found that for every degree of increased temperature the pressure increased a uniform amount. This uniform amount meant the pressure of a gas will increase about 1/273 of its pressure at zero degrees Celsius for every degree Celsius rise in temperature if the volume is kept constant. Thus a closed tank full of gas will have a higher pressure when it has an elevated temperature, and as the temperature decreases the pressure will decrease.

In 1802 another Frenchman, Gay-Lussac found that gas will expand at a uniform rate as the temperature increases. The law of Gay-Lussac means that the volume of every gas will increase 1/273 of its volume at zero degrees Celsius for every degree Celsius rise in temperature, if the pressure is kept constant.

THE GAS LAWS

In summary - Robert Boyle explains the effects of compressing a gas, stating that the pressure which a given quantity of gas at constant temperature exerts against the walls of the containing vessel is inversely proportional to the volume occupied.

Jacques Charles explains that the pressure of every gas will increase 1/273 of its pressure at zero degree Celsius for every degree Celsius rise in temperature if the volume is kept constant.

Gay-Lussac explains that the volume of a given weight of every gas will increase 1/273 of its volume at zero for every degree Celsius rise in temperature, if the pressure is kept constant.

PASCAL'S CONCLUSION

The name Pascal may bring up the thought of power hydraulic principles. Pascal's conclusion states that pressure applied anywhere to a body of confined *liquid* is transmitted undiminished to every portion of the surface of the containing vessel. We recognize that compressed air also follows the teachings of Pascal. Static pressure within a typical pneumatic system is equal at all surfaces. If a system is very long vertically, the system pressure depends upon the height of the air column. Practically, this is usually not of concern in typical pneumatic power transmission systems.

BERNOULLI'S PRINCIPLE

Force water through a constricted pipe with pressure gauges at *a*, *b*, and *c* (Figure 3.10). The pressure will be lower at *b*, higher at *a* and *c*.

Bernoulli's Principle states that *there is in general such a relation between pressure and velocity, and that whenever the velocity of a fluid (like air or water) is high the pressure is low, and vice versa.* The special shape of an airplane wing, with its curvature, smooth leading edge, and pointed trailing edge has been developed because it makes the air travel faster over the top surface than over the bottom (Figure 3.11), where the length of the arrows is proportional to the speed of the air. Thus, by Bernoulli's Principle the pressure over the upper surface is low and that over the lower surface is high, providing the necessary lift for an airplane. These principles have found practical application in many material handling systems.

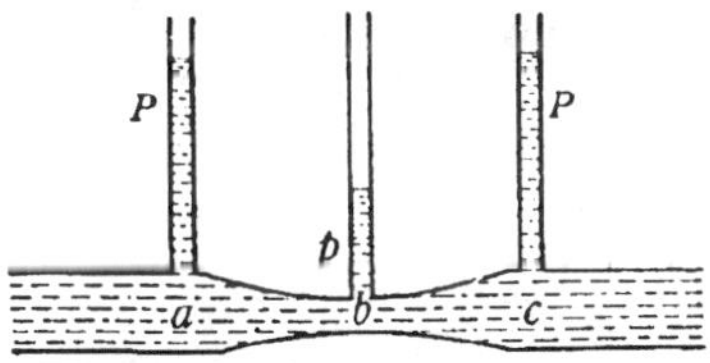

Figure 3.10 Bernoulli's Principle proves pressure is less at (*b*) where speed is greater than at (*a*) or (*c*).

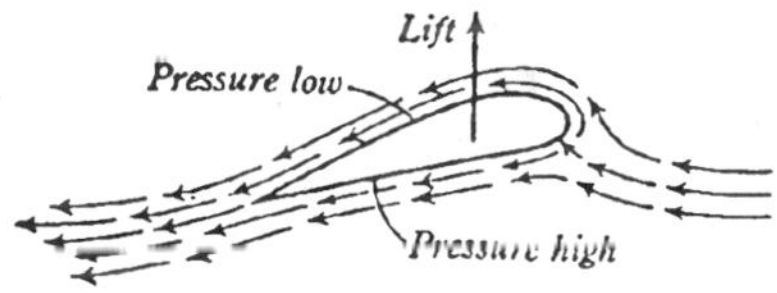

Figure 3.11 The lift on an airplane can be explained by Bernoulli's Principle.

PNEUMATIC EJECTION SYSTEMS

Knowledge of the characteristics of moving air is helpful in creating an effective ejection system for plastics, die castings and other movement of parts and products. Careful analysis of the moving air has made the use of pneumatics an essential in the graphic arts areas. Positioning of plastic parts for printing and packaging has involved large amounts of pneumatic equipment.

SUMMARY

Force and motion design parameters can be established for pneumatically-powering machines with the predictability ensured by the proven gas laws. A knowledge of the gas laws aids in troubleshooting a machine malfunction. Changes in pressure or temperature can be easily related to machine malfunctions.

REVIEW QUESTIONS

1. What is a Torricellian vacuum?
2. How is sea level involved with pneumatic pressure measurements?

3. What is the basic content of the law named after J. A. C. Charles?
4. What is the meaning of the term “relative humidity”?
5. Explain “standard air” and “free air”.
6. Why has Boyle’s Law become essential to explain the “pneumatic spring” phenomena?
7. What type of gas would typically be employed in an aircraft accumulator?
8. What type of gas would typically be employed in a machine tool accumulator?
9. How does Pascal’s conclusion relate to pneumatic systems?
10. How can a knowledge of the gas laws aid in the troubleshooting of a machine malfunction?

4

FORCE AND MOTION

INITIATION AND CONTROL OF FORCE AND MOTION

THE POWER FUNCTION - THE MOTION FUNCTION

Use of compressed air or vacuum to initiate and control force and motion offers some significant advantages. When compared with other power transmission systems it also has some limitations.

THE POWER FUNCTION

Force values for a compressed air power device are calculated by the product of the effective surface area of a piston face or the effective displacement of a rotary device and the applied pressure level.

As an example we can calculate the number of square inches exposed to the compressed air when using a typical pneumatic cylinder by multiplying the radius squared times 3.1416. The answer will be in square inches. To determine the force it is only necessary to multiply the pounds per square inch by the number of square inches of the piston face. This neglects the friction that will be encountered. The rod area can also be calculated. If the cylinder is to be used in a typical pull mode the effective area will be the piston area minus the rod area.

Force values for a vacuum power device are calculated by the effective surface area of a piston face or the effective displacement of a rotary device and the difference in pressure between ambient atmospheric pressure on one face and a reduced pressure on the opposing face created by the action of the vacuum pump.

THE MOTION FUNCTION

Many factors affect the rate at which an air cylinder will move. The circuit and information shown in Figure 4.1 provides data relative to rate of movement for simple operations. As mentioned in Figure 4.1, precision control of the rate of movement of a pneumatic cylinder is not a function of the rate at which compressed air is supplied to the actuator. Effective control of speed with hydraulic devices with a pneumatic cylinder to provide the basic power can provide an inexpensive way to provide power and control of small drill presses, milling machines, etc. The commercial term "Hydro-Check" refers to a hydraulic cylinder assembly fitted with a flow control device to meter oil from one end of the hydraulic cylinder to the other at an adjustable controlled rate. This controls the rate of movement of the mechanical device which is being powered by the pneumatic cylinder.

Vacuum system actuators move at a rate commensurate with the rate at which air is exhausted. Restriction of atmospheric pressure to the actuator input can reduce the rate of movement.

THE PNEUMATIC ADVANTAGE

Two major uses enhanced the value of compressed air systems in our industrial sector. The first was the air brake for railroad rolling stock, invented and promoted by George Westinghouse. The second was the pneumatic tire, which became the major interface between the machine and the roadway, on which vehicles of all types depended, especially the automobile.

The air brake, designed for use with railroad rolling stock, provided a "fail safe" alternative to the various mechanical devices used early in railroad history. An entire train could be controlled by compressed air from one or more points of control. With a pneumatic system, braking is relatively uniform from one end of the train to the other. The cost of the compressed air was not of great concern with the massive power already available in the locomotive. The principle employed in the railroad air

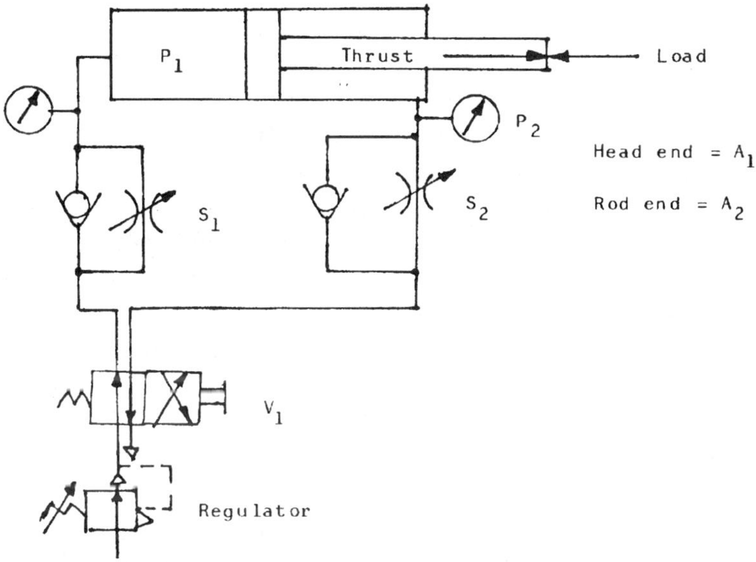

Figure 4.1 Conventional "speed control" of a pneumatic cylinder. Meter-out - Thrust = $P_1A_1 - P_2A_2$. If load stays constant, traverse velocity remains constant if P_1 stays constant. A regulator ahead of the valve V_1 will suffice. *But* - if the load changes - for *any* reason - the thrust changes. P_2 must change because the regulator keeps P_1 constant. If P_2 changes, the air flow rate across S_2 changes as air flow depends on Pressure P_2. S_1 and S_2 are not very effective in controlling speed. How *can* speed be controlled? By using a hydraulic motion controller coupled to the machine "Hydro-Check™" - or - by controlling the pressure at the adjustable orifice of the "speed control" valves. This can be done by sensing the pressures P_2 at S_2 and using that pressure as the control to drive a regulator controlling P_1. Thus - if load drops, P_2 will increase - unless P_1 is dropped. The devices to accomplish this are available now. Only modifications are needed. Trade Mark - Bellows Division Parker Hannifin Corporation.

brake soon migrated to other industrial power needs and the pneumatic power industry was born.

The automobile tire created a need for a quick method of inflation. Hand and foot operated air pumps were slow and labor intensive. The relatively small, compact air compressors designed for automotive service stations provided an answer for inflating tires. They soon blossomed into an essential power source for car hoists, lubrication pumps to develop pressures sufficient to pump grease at pressures ranging from 3,000 to 12,000 psig, and to drive various power tools. The expanding use of compressed air for servicing functions in the automotive area changed many portable installations into central air supply systems. Major tools emerged such as powered tire change machines, exhaust system pipe benders and special air-powered machinery such as frame-straightening mechanisms. The air-powered wrenches and nut runners are considered an essential tool in most service areas. Industrial uses of pneumatic and vacuum power included packaging, ejection mechanisms and other mechanical motions where speed and resiliency are essential.

The ability to intensify air pressure by using a relatively large air cylinder as an actuator to power a relatively small diameter piston to move grease makes it possible to provide pressures on the grease from 3,000 to 12,000 psi. In some applications pressures to 30,000 psi can be developed as needed. This type of application is found in most every commercial garage and service station.

Compressed air may be used in areas where a comparable hydraulic system may be smaller and more efficient. The compressed air system, however, may be a better choice because of potential fire or explosion. The pneumatic system is relatively fire safe. Pneumatics are a first choice in the operation of large gas valves where explosions are an expected hazard. The possibility of an explosion in grain elevators, flour mills and other operations where flammable powders exist also makes air power and/or control a good choice.

Cleanliness may also be a significant reason for using compressed air power. Food, paper, textiles and comparable materials can be readily damaged by even a minor oil leak from a hydraulic system. Some systems use equipment which is specifically designed for "dry" air. Lubrication needs are minimized or eliminated by the use of low friction materials and special packings.

THE PNEUMATIC DISADVANTAGE OR LIMITATION

Most industrial pneumatic systems function with pressure levels at 150 psi or less. Typical industrial pressure levels are from 60 to 100 psi. Automotive service stations can need up to 150/200 psi for pneumatic tires for heavy mobile equipment with 30 psi being typical for automotive tires. This industrial and service pressure level limits the practical force capabilities of a compressed air system. Vacuum systems are limited to the capabilities of the vacuum pump to exhaust air and the nominal atmospheric pressure at the point of power usage.

Hydraulic systems using a liquid in place of a gas may offer significant economies in the basic linear and rotary power functions. The hydraulic system must have a reservoir of significant size and associated return lines. The pneumatic system may be required for other functions so that for intermittent service the use of less efficient tools can be easily justified.

Thus, a pneumatic system which may have a power transmission cost five times that of an equivalent hydraulic system can be justified because of other pneumatic requirements in the immediate area. First cost of a hydraulic system may be significantly greater than a pneumatic system that can be supplied from an existing compressed air source and distribution network.

SUMMARY

Pneumatic flow control devices are affected by the compressibility of the air. Hydraulic devices powered by compressed air may be a preferred system.

Need for inflating the pneumatic tire may be the economic justification for many pneumatically-powered tools.

Cost analysis needs to include safety, portability, first cost versus operating costs and sophistication of force and motion requirements.

REVIEW QUESTIONS

1. Can pneumatic systems employ meter-out technology efficiently?
2. What is the primary advantage of pneumatic systems in hazardous locations?
3. Name a major consideration in employing pneumatics for powering portable tools.
4. What are typical operating pressures for a pneumatic system?
5. Why are pneumatic systems a primary choice for food and paper processing machinery?
6. What was the pneumatic advantage that resulted in major applications for railroad brakes?
7. How is air pressure employed to power grease guns?
8. What is a typical pressure for an automobile tire?
9. What is a typical pressure for equipment employed on mobile equipment?
10. Why does an explosive atmosphere dictate use of compressed air for powering force and motion devices?

5

THE POINT OF USAGE

MOVING MATERIALS

A March wind is an excellent example of moving materials with reckless abandon (Figure 5.1). The tropical hurricane is an awesome example of moving air and its inherent power (Figure 5.2). Similar forces programmed on a minute scale serve to accomplish household and industrial tasks (Figure 5.3). In the distant past it was the major source of power for moving freight and passengers over the world's waterways (Figure 5.4). Today, it is a vacation get-away from the tiny surf sailors to the magnificent sailing yacht (Figure 5.5).

CAPTURING THE POWER

The Egyptians and Chinese were probably the earliest known pioneers in the use of wind to provide power for grinding grain, pumping water and myriad tasks where this "free" power was used to lessen the burden of farm animals and their masters (Figure 5.6). The square vertical drive provided a natural source of power to activate the shaker assembly fastened to the grain hopper. A light wind would rotate the vertical drive slowly and the grain flowed slowly to the mill stones. As the speed increased with a stronger wind the increased speed caused the hopper to vibrate at a faster rate and increased the supply to the mill stones. The Europeans at a somewhat later date also pioneered in some of the exten-

Figure 5.1 The wind is nature's broom.

Figure 5.2 The hurricane is an uncontrollable power.

Figure 5.3 The blower is great for removing grass clippings from the driveway.

Figure 5.4 The wind was harnessed with crude fabrics and rope to provide the power for moving passengers and freight over natural waterways. Man learned to employ his brain power to supplement his muscle power.

Figure 5.5 The sleekly-designed sailing craft is often the last of the "quiet" vacation retreats.

a *b*

Figure 5.6 The windmill was essential to the pre-steam and electricity era. (*a*) Angle pin-and-socket gearing of hardwood transferred the energy from the rotating blade to the vertical drive for the mill stones. (*b*) The square mill stone drive pole vibrated the grain hopper at a rate commensurate with the wind velocity, preventing grain pile-up and course flour. An ancestor to our modern automation systems.

sive uses of wind power to supplement that of man and beast. Prior to the availability of rural electric power, the windmill was a common sight from coast to coast and border to border in North America (Figure 5.7). There is still considerable interest in using the windmill to generate power in remote locations (Figure 5.8).

Figure 5.7 The angle-iron-frame windmill of the plains is fast disappearing in North America.

Figure 5.8 The modern windmill is streamlined and efficient.

ENJOYING THE POWER

Youngsters (and often adults) enjoy using the wind to fly kites (Figure 5.9) and revel in the freedom associated with hang gliding and the more sophisticated, not-powered glider or sail plane (Figure 5.10). Birds soaring on the wind currents have always interested the poet and story teller (Figure 5.11).

REVERSING THE FLOW

The awesome vacuum created by tornadoes is well known to those in the plains states (Figure 5.12). The benevolent vacuum created by the household cleaner has been a boon to housewives and industrialists alike (Figure 5.13). Blowing air into a container which will allow the air to pass through with relative freedom and capture the particulate materials in the airstream has almost obsoleted the spring cleaning rug beater (Figure 5.14). Of course it will not capture vapor, moisture, etc. The so-called

Figure 5.9 Children's kites demonstrate wind power and the dynamics of flight.

Figure 5.10 Unpowered flight is quiet and sometimes scary.

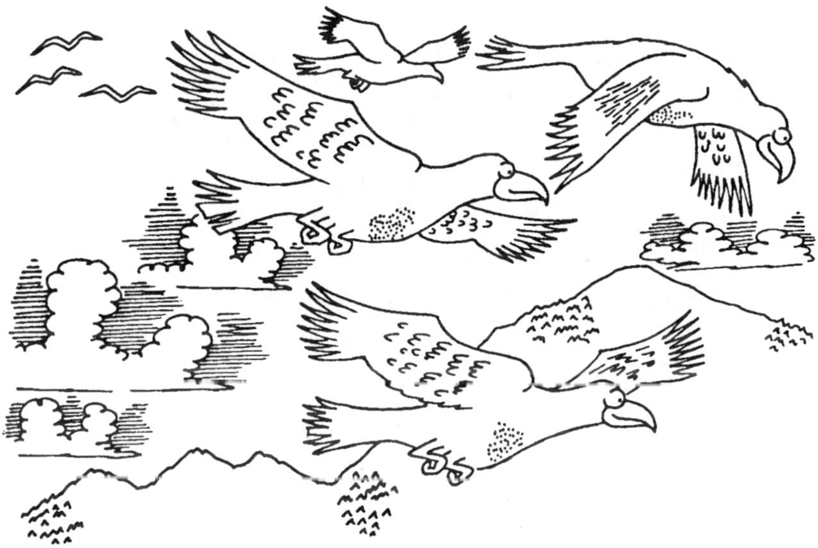

Figure 5.11 The birds understand efficient use of the air currents.

Figure 5.12 A tornado can create a powerful vacuum.

Figure 5.13 No house should be without a vacuum cleaner.

a

b

Figure 5.14 (*a*). The annual rug beating was rough on the rug and the person wielding the beater. (*b*). The essential domestic vacuum cleaner bag catches the debris.

"wet-dry" vacuum cleaner designed with a tank that will receive either dust or liquid is the exception. The motor and moving elements are located in the upper portion of the receiver tank, and provisions are included to avoid entry of water or solids into the working mechanism. Within industry it has been a prelude to cleaner working areas and a safer environment (Figure 5.15). Virtually every town has its powered street sweeper and vacuum to remove autumn leaves, winter dirt and the windblown debris (Figure 5.16).

THE POWER ASSOCIATED WITH INFLATION

Inflating the pneumatic tire requires power (Figure 5.17). In this instance the air is employed as the resilient fluid to maintain the tire in a pre-determined shape within the casing. The rough road surface or itinerant stone can be accommodated by easily deflecting the casing without major movement of the vehicle. The air compressor needed to accomplish this task with electric power (Figure 5.18) is also available to provide the pneumatic power to perform other maintenance tasks (Figure 5.19).

Figure 5.15 The industrial sweeper makes for a neater and safer working area.

Figure 5.16 Clean streets are available to the smallest town.

Figure 5.17 The hand pump had its limitations.

Figure 5.18 The electric-powered compressor is a labor saver.

Figure 5.19 The pneumatic power accomplishes other tasks with ease.

THE POWER ASSOCIATED WITH DEFLATION

An industry has been built around the ability to remove air from one half of a die with a formable plastic sheet between the die halves. Vacuum forming is used in the manufacture of such varied devices as refrigerator door liners, automotive trim, toys and many other everyday products found in the home and industry. With vacuum cups it is possible to move massive loads or minute parts in assembly procedures or material processing (Figure 5.20).

POWER TRANSMISSION VIA CONTAINED AIR

Compressed air directed to one port of a pneumatic cylinder or to one port of a pneumatic motor will cause the cylinder rod to move or the motor shaft to rotate, if the air pressure can overcome the static friction and/or the load imposed on the cylinder rod or motor shaft (Figure 5.21). Likewise, atmospheric pressure will cause a cylinder rod to move or a vacuum motor shaft to rotate if the difference in pressure is great enough as the vacuum pump reduces the pressure at the opposite port of the motor or cylinder. The linear movement may be accomplished by contained pressure in a bag or diaphragm structure in place of a cylinder and rod (Figure 5.22). Limited rotation can be accomplished by many different types of rack and pinion assemblies (Figure 5.23) or limited rotation vane

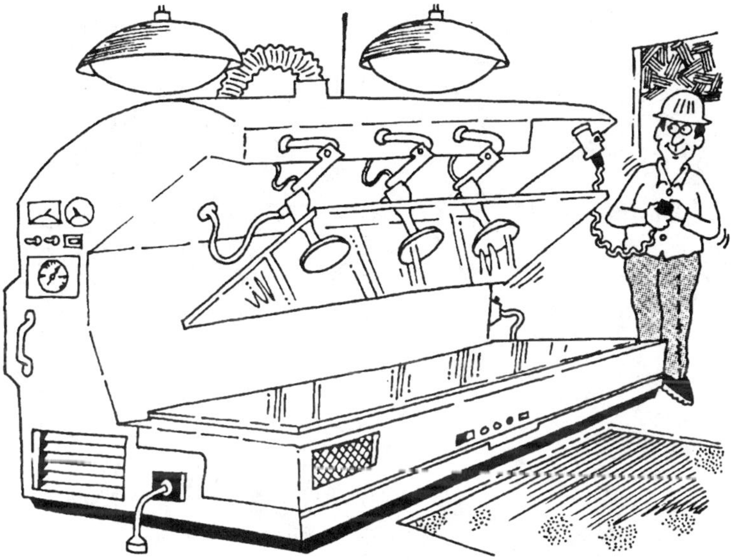

Figure 5.20 The robot can easily handle sheet glass with a vacuum gripper.

a

b

Figure 5.21 Air pressure times piston area must be greater than load plus friction. (*a*). Linear push. (*b*). Rotary force.

Figure 5.22 An air pressure container can be a flexible elastomeric bag. Various shapes are available so that change of shape may be of greater significance than expansion capabilities.

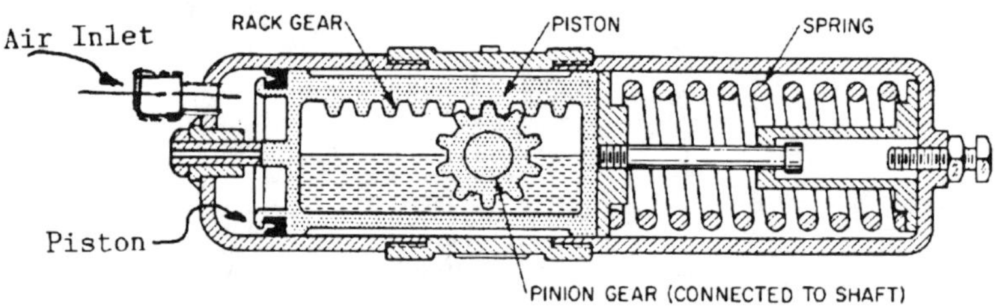

Figure 5.23 A rack and pinion assembly can provide rotary motion.

structures (Figure 5.24). Two simple ram-type cylinders may push against a lever structure to cause limited rotation (Figure 5.25).

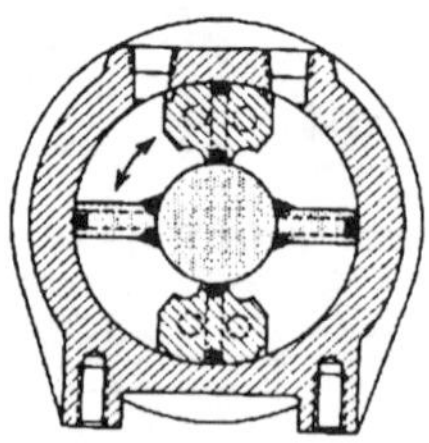

Figure 5.24 Vane-type assemblies take little physical space.

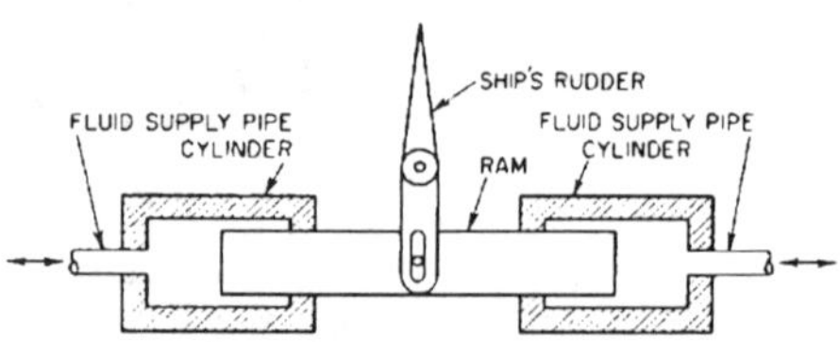

Figure 5.25 Two simple ram-type assemblies can provide a limited rotational force.

FLOATING ON AIR

Massive loads can be moved on a cushion of compressed air. Specially designed vehicles provide a continuing supply of compressed air between the vessel and the water or land over which it hovers so that it is held at a pre-determined distance from the static surface. The craft can then employ a propeller or jets for linear horizontal movement. Heavy machinery can be moved in a similar manner with a supply of compressed air under the machine as it is pushed or pulled to the desired location (Figure 5.26).

Figure 5.26 Compressed air between large surfaces can hold heavy loads off the floor at pre-determined distances for ease of movement.

LINEAR FORCE AND MOTION

Properly applied, pneumatic cylinders can provide outstanding performance as actuators in a variety of automatic machines and devices. Advantages of pneumatic cylinders in this type of service stem mainly from their relative flexibility of application, large force-exerting capabilities, high power per unit weight, size and low cost.

Widespread and growing applications of pneumatic cylinders in industry suggests that a review of basic selection and application criteria would interest many manufacturing engineers in an effort to minimize downtime and provide the best possible force and motion functions.

Accordingly, this Chapter Five covers some of the important factors that should be considered in specifying, installing and maintaining pneumatic cylinders. Although linear-motion pneumatic cylinders will be discussed, many of the factors addressed apply to rotary pneumatic actuators as well. Some aspects of hydraulic cylinders will also be addressed because of the common problems that can occur in both hydraulic and pneumatic cylinders. Relatively large rods may be important in a pneumatic cylinder for needed column strength. The same rod may be important for column strength in a hydraulic cylinder and also because of a desire for additional speed by the use of a regenerative circuit with the large rod hydraulic cylinder. The regenerative circuit is usually considered as a hydraulic option not practical for pneumatic usage.

BASIC CYLINDER DESIGN

The choice of a supplier for a pneumatic cylinder may largely determine the unit basic design. Some dimensional standards have been developed in the fluid power industry relative to mountings, rod-ends, port sizes, etc. It is at this point in the selection stage that the envelope and mounting dimensions are set and interchangeability with cylinders made by other manufacturers is established (Figure 5.27). The choice of supplier may also fix the type of cylinder construction which will be used.

CONSTRUCTION FEATURES SHOULD BE CAREFULLY CONSIDERED

Construction features should be carefully considered in a particular application. For example, the method of fastening the cylinder heads to the cylinder body is closely related to physical design of the end product. As to gland design, the quality control engineer should verify that the gland will automatically provide a drip-proof seal without the need for periodic adjustment. The effective piston-bearing length should be adequate for the job, and the joint between the piston and piston rod should have the necessary strength and means to prevent loosening of the piston and rod joint (Figure 5.28). This is especially important for applications where the piston may stop with an impact against the ends. Materials, clearances and finishes used in fabricating the cylinder should meet any special requirements of the application.

If the application is in an area where fine particulate matter may be present (as in a foundry), the bearing surfaces should be designed to pro-

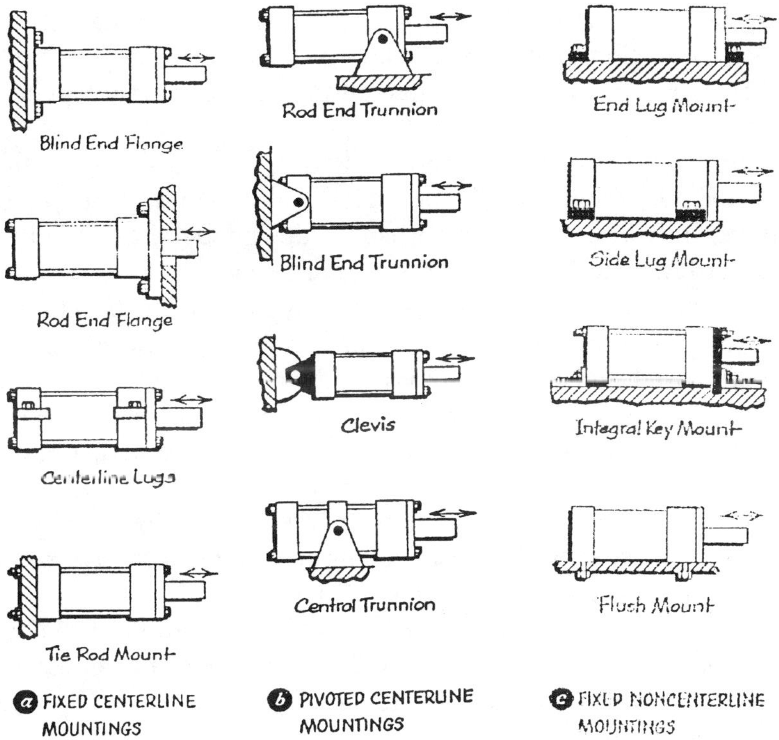

Figure 5.27 Pneumatic cylinders are available with a variety of mounting arrangements. Fixed centerline mountings (*a*) and pivoted centerline mountings (*b*) provide for piston rod thrust to be taken up along cylinder centerlines. This minimizes swaying of the cylinder and flexing of the machine members to which the cylinder is mounted. Cylinders with fixed non-centerline mountings (*c*) are convenient in many applications where thrust may be taken up at offset mounting surfaces.

vide means of minimizing wear. As an example, surface B in Figure 5.28*a* must be significantly smaller in diameter than surface E so that the resilient seal rather than surface B can scrape the foreign matter off and protect the *major* bearing surface at E. Suitable retaining spacers can hold seal C shown in Figure 5.28*b*. The two seals "C" can protect bearing surface D from foreign matter in either direction of movement.

PRESSURE RATINGS

Unfortunately there is no standard method of establishing pressure ratings for industrial cylinders. Some manufacturers rate their cylinders on the basis of ultimate strength; others publish pressure ratings based on a typical yield strength or fatigue strength. There is also no consensus on a design factor. Some cylinder designers typically use a design factor as low as 2 to 1 based on minimum expected strength, and others specify 3-1/2 or 4 to 1 on yield.

Pressure ratings for pneumatic cylinders may be of little significance if the cylinder line is likely to be used for both hydraulic and pneumatic service. Industrial cylinders are usually designed to have a unit tensile stress of 12,000 to 15,000 psi in the cylinder body at rated working pressures. Since the cylinder body is generally made from low carbon, cold drawn steel tubing which typically has a tensile yield strength of about 55,000 psi, a design factor of approximately 4 to 1 on yield is provided at rated pressures. This is also the minimum design factor specified by several automobile manufacturers for the standardized, heavy duty automation cylinders. In many portions of these components the design factor is much higher. As an example, hydraulic cylinders currently being designed for use with missile launching equipment also have a specified design safety factor of 4 to 1 based on yield strength.

Dimensional changes occurring in stressed cylinders may further limit the allowable working pressures since large dimensional changes can seriously affect the performance and life expectancy of non-metallic seals (Figure 5.28).

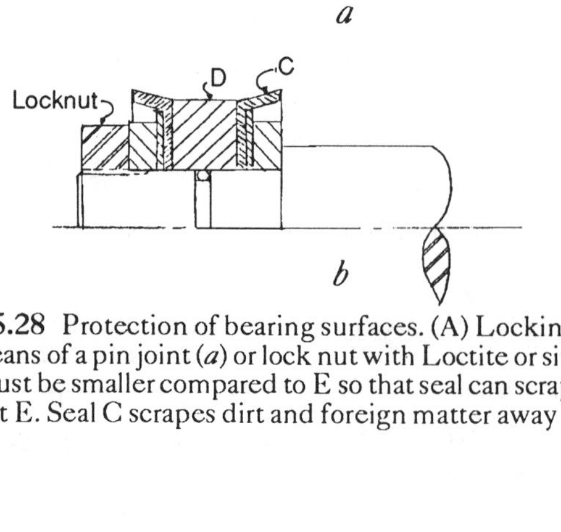

Figure 5.28 Protection of bearing surfaces. (A) Locking rod to piston may be by means of a pin joint (*a*) or lock nut with Loctite or similar sealer (*b*). Surface B must be smaller compared to E so that seal can scrape dirt from bearing surface at E. Seal C scrapes dirt and foreign matter away from surface D.

Generally, if the published pressure rating of a cylinder is based on too low a design factor or results in excessive unit stresses, operation at that pressure may cause unsatisfactory performance, failure, or a short useful life. If the circuit in which a cylinder is to be used can impose high shock pressure in the cylinder, additional derating will be required.

WORKING MEDIUM

When air is used as the power source, corrosion of the cylinder bores may be a problem. In this case, steel cylinder bores should be plated to attain corrosion resistant surfaces, or inherently corrosion-resistant materials (brass or stainless steel) should be specified. Normally, provisions should also be made for positively lubricating the cylinder during operation. In general, seal manufacturers and designers tend to agree that seals made from Buna N and similar elastomers provide the most efficient, cost effective sealing performance on lubricated shafts.

ASSEMBLY OF TIE ROD-TYPE CYLINDER IS CRUCIAL

Surfaces adjacent to the cylinder tube must be maintained in a parallel position, especially in sizes of approximately 4-inch diameter and larger (Figure 5.29). Tie rods must be tensioned uniformly - or heads will not be parallel. Problems that may be difficult to analyze occur as the piston comes to rest against heads at the end of stroke. Seal and wear problems shorten life.

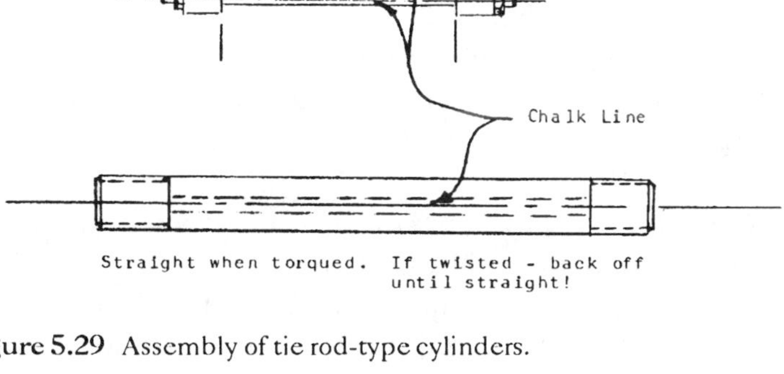

Figure 5.29 Assembly of tie rod-type cylinders.

One quick way of checking - tap the tie rods - they should "ring" at about the same frequency as judged by ear. Also, tie rods should not be twisted. Check by making a chalk line on tie rods before torquing. The line should not be spiral when done! If a rod is not straight, back-off slightly until the line is straight. If you do not have straight tie rods - cylinder heads may wind up "twisted" after the cylinder is pressurized a few times (Figure 5.30).

Heads twist if tie rods are not straight.

Figure 5.30 Absence of straight tie rods - cylinder head may end up "twisted" after cylinder is pressurized several times.

CYLINDER MOUNTING TYPES

Assuming that the basic pneumatic cylinder design and pressure rating has been established, the next step is to select the cylinder mounting type that will be best suited to the application. This choice involves consideration of the various mounting types available, which may be classified generally by whether or not thrust is absorbed on the cylinder centerline. Mountings where this action does occur are available in fixed mount styles and pivot mount styles (Figure 5.27*a*, and *b*). In mounting types where thrust is not absorbed on the cylinder centerline (Figure 5.27*c*), the thrust from the piston rod is taken up at a mounting surface that is usually parallel with, but not coincident with, the cylinder centerline.

MOUNTING SPECIFICATIONS

Specification of the type of mounting for a cylinder depends largely upon the application. *Some of the factors that should be considered include:*

1. Cylinders with non-centerline-type mountings tend to sway under load (Figure 5.31*a*). Resulting additional wear on the cylinder can shorten the useful life of the component. This is especially important on larger bore cylinders.
2. The rigidity of the machine frame should be taken into account. It should be remembered that cylinders with non-centerline-

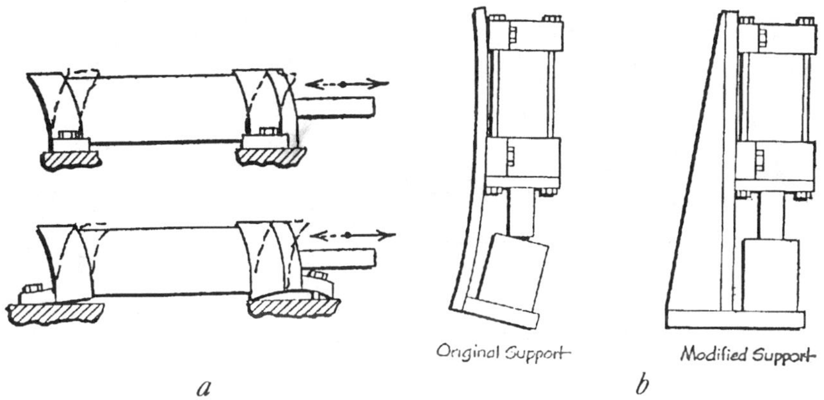

Figure 5.31 Non-centerline-type mountings. (*a*) In applications involving large forces, cylinders with non-centerline-type mountings tend to sway under load. (*b*) Use of non-centerline-type cylinder mountings may require strengthening machine members to resist bending under load. In large cylinders over a 6-inch diameter, swaying can create premature failure of seals - both static and dynamic.

type mountings often require stronger machine members to resist bending moments (Figure 5.31*b*).

3. If the motion of the machine member acted upon by the piston rod is essentially linear, then in most cases a fixed mounting-type cylinder should be used. If the machine member moved by the piston rod travels in a curvilinear path, a pivot-mounted cylinder is the obvious choice. Pivot mountings are available with the pivot points on the rod-end, on the blind end, or centrally located near the mid-point of the cylinder body. In most cases, a layout of the rod end path will determine the best pivot mounting to be used.
4. The matter of cylinder strength versus stroke length should be considered. Generally, long-stroke, pivot-mounted cylinders of the center trunnion type can use smaller rod diameters without danger of buckling than can similar cylinders with the pivot at the blind end. This can be verified by reference to stroke or buckling charts published by many cylinder manufacturers. Where long cylinders employ fixed mountings, additional supports are often required to prevent excessive sag or vibration (Figure 5.32). Fixed non-centerline-mounted cylinders with very short stroke lengths introduce another type of cylinder

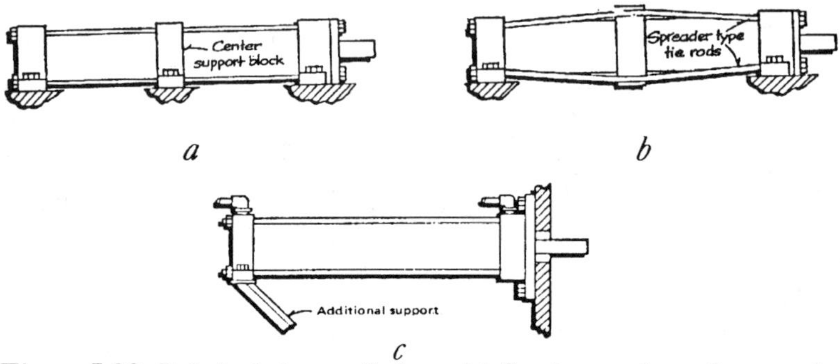

Figure 5.32 Relatively long cylinders with fixed mountings often require additional supports to prevent excessive sag or vibration. Arrangement at (*a*) uses an extra mounting block located midway along cylinder body. At (*b*), spreader-type tie rods are used to increase rigidity in center portion of cylinder. Where one end of a cylinder must be overhung, as at (*c*), an additional supporting member can be provided. Care must be used in tightening tie rods as this puts column loading on the cylinder. This is worst at 0 psi. Column loads may buckle the cylinder.

problem (Figure 5.33). In this case, mounting bolts are subjected to increased tension which, in combination with shear forces can overstress the bolts.

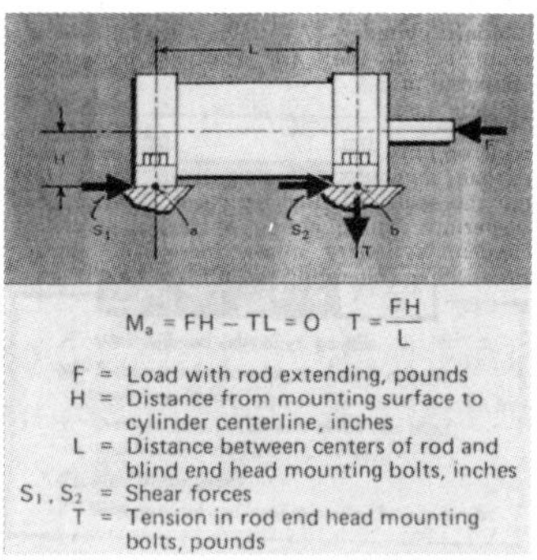

$$M_a = FH - TL = 0 \quad T = \frac{FH}{L}$$

F = Load with rod extending, pounds
H = Distance from mounting surface to cylinder centerline, inches
L = Distance between centers of rod and blind end head mounting bolts, inches
S_1, S_2 = Shear forces
T = Tension in rod end head mounting bolts, pounds

Figure 5.33 Relatively short, fixed non-centerline-mounted cylinders can subject mounting bolts to large tension forces. This, in combination with shear forces, overstresses the bolts. Condition shown for a cylinder rod-extending against a load illustrates manner in which tension is developed in rod end head mounting bolts. Note from the formula that for a constant load, tension increases with decrease in distance L between rod and blind-end head mounting bolt centers. Similar analysis can be applied to determine stressing of blind end head mounting bolts by substituting for load F the rod tension force developed during retraction, and by equating moments about point b to zero. In some cases, where large bore cylinders have a thin wall cylinder tube, the bolt loading and internal pressure may cause the cylinder body to enlarge in diameter near the center of its length. This can cause piston seal problems and fatigue the cylinder wall.

FORCES IN TENSION OR COMPRESSION

Selection of a mounting style also depends upon whether the major force applied to the machine will result in tension or compression of the cylinder rod. The blind end, flange-type mounting is best for thrust loads (rod in compression), and a rod end flange mounting is best where the rod is stressed in tension.

POTENTIAL ALIGNMENT PROBLEMS

Alignment problems should always be considered. If misalignment can occur between a cylinder and the machine member it moves, it may be necessary or desirable to compensate for this in the selection of the cylinder mounting. For example, a simple pivoted centerline mounting will compensate for misalignment if the misalignment is primarily in one plane. Typical mountings of this type include clevis and trunnion arrangements. Where misalignment can occur in multiple planes, then the cylinder should be equipped with self-aligning ball joints both at the blind end of a clevis-mounted cylinder and in the rod end member.

NOTES ON INSTALLATION

Proper installation of a cylinder can be controlled to a considerable extent by the engineer when a machine layout is made. Drawing notes clarifying methods of installation; precautions to be observed, etc., are helpful to installation personnel. *Some points that manufacturing engineers should consider in the layout stage include:*

1. If high shock loads are expected, the cylinder should be mounted so that full advantage is taken of the cylinder's elasticity. Using the inherent elasticity of a cylinder in shock situations can mean the difference between a successful application and one that is always troublesome. High shock conditions usually require more rugged fastening provisions than are normally employed.
2. Fixed mounting cylinders should be held in place by keying or pinning, and some provisions should be made for this (Figure 5.34). Cylinders with integral key mounts may be used where keyways can be cut in a machine member. This type of mounting takes up shear loads, provides accurate alignment of the cylinder, and simplifies installation and servicing.
3. The use of separate keys to take shear loads is relatively common. Shear keys should be placed at the proper end of a

cylinder-at the rod end if major shock load is in thrust, and at the blind end if major shock loads are in tension (Figure 5.34). In a given installation, only one end of a cylinder should be keyed to the machine. If both ends are keyed, the advantages of cylinder elasticity in absorbing shock will be lost. Temperature and pressure effects should also be considered. A cylinder that is mounted at room temperature and zero pressure may be subjected to working temperatures ranging from 120 to 140 degrees F or more, and pressures of 2,000 or 3,000 psi if it is a hydraulic cylinder. Under these operating conditions it is obvious the cylinder length will increase over its installed length. The cylinder should *never* be mounted so that it is not free to expand and contract. For an air cylinder - 18-inch stroke length, brass body - will elongate 0.021 inches for a 100 degree F temperature rise. Tie rods will elongate about 0.015 for the same use, depending upon mount, which could be a problem source. Even without the presence of serious shock, temperature and pressure effects, use of shear keys at both ends of a cylinder is *not* advisable. The difficulty of manufacturing a cylinder to an exact over-all length so it will fit between keys located for another cylinder is the reason.

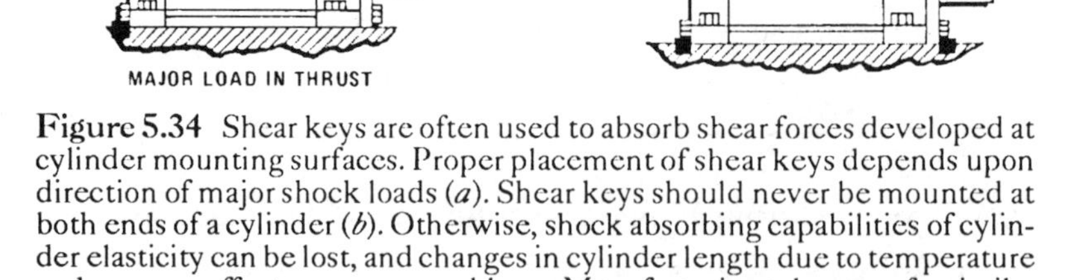

Figure 5.34 Shear keys are often used to absorb shear forces developed at cylinder mounting surfaces. Proper placement of shear keys depends upon direction of major shock loads (*a*). Shear keys should never be mounted at both ends of a cylinder (*b*). Otherwise, shock absorbing capabilities of cylinder elasticity can be lost, and changes in cylinder length due to temperature and pressure effects can cause problems. Manufacturing tolerances for similar components could also make replacement difficult with such arrangements.

4. Locating pins may be used instead of shear keys to help absorb shear loads and to obtain cylinder alignment (Figure 5.35*a*). Cylinders should be pinned at either end (and not both ends) for the

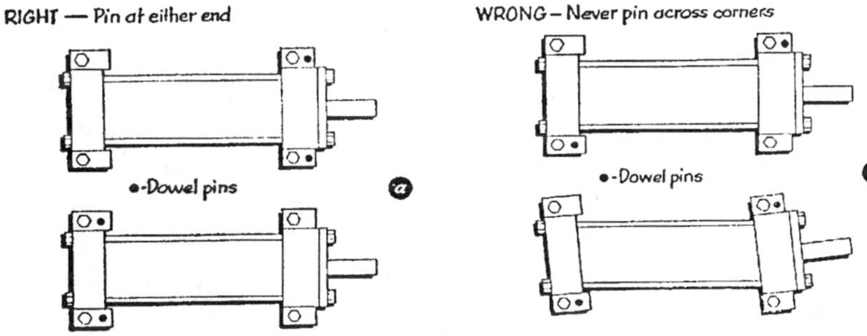

Figure 5.35 Cylinders pinned in place to help secure alignment and resist shock loads should be pinned at either end, but not at both ends (*a*). Choice of end depends upon direction of major shock load. If dowel pins are used across corners (*b*), cylinder may be warped by operating temperatures, pressure or shock loads.

reasons mentioned previously. Occasionally a designer will carry over die design experience and call for pinning across corners, as in Figure 5.35*b*. This can result in severe warping of a cylinder when it is subjected to operating temperatures and pressures.

5. Pivoted mounts should have the same type of pivots at the cylinder body and the rod end. If a simple pivot pin mount is used, the pivot pin axes at each end should be parallel. If a ball joint or self-aligning bearing is used at one end, a similar device should be used at the other. Trunnion mounts are generally designed to resist only shear loads at full-rated pressure. Therefore, self-aligning mounts should not be used to support the trunnions if the cylinder is to be used near full-rated pressure, since bending forces can also be set-up (Figure 5.36).

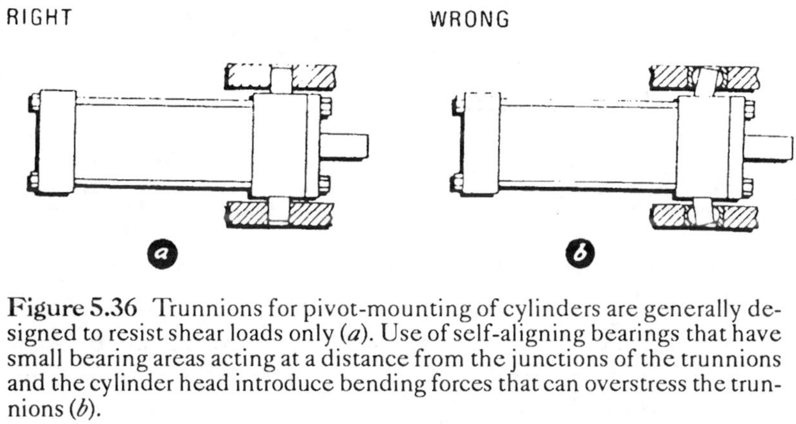

Figure 5.36 Trunnions for pivot-mounting of cylinders are generally designed to resist shear loads only (*a*). Use of self-aligning bearings that have small bearing areas acting at a distance from the junctions of the trunnions and the cylinder head introduce bending forces that can overstress the trunnions (*b*).

6. Mis-alignment that can occur when a fixed mount cylinder is located so that its centerline does not coincide with that of the actuated machine member is of two types, as in Figure 5.37. While such a cylinder may tolerate misalignment that increases with the stroke, it cannot tolerate misalignment which remains constant throughout the stroke. Section 6.3.2 of the Joint Industry Conference (JIC) hydraulic standards states "Cylinders shall be located in alignment with work slides, and shall be such that no side or radial load shall occur on piston rod or ram, unless other suitable provisions are made to take such loads". This recommendation is a good rule to follow in mounting a cylinder. In some cases, misalignment is inevitable. When this occurs, an expedient is to mount the cylinder in the manner shown in Figure 5.38. With this arrangement, oversize holes in the side lugs of a rod end cylinder head allow the rod end of the cylinder to move up and down with respect to a pair of dowel pins. The blind end head is fastened rigidly to the machine, and flexure is provided by the cylinder body.

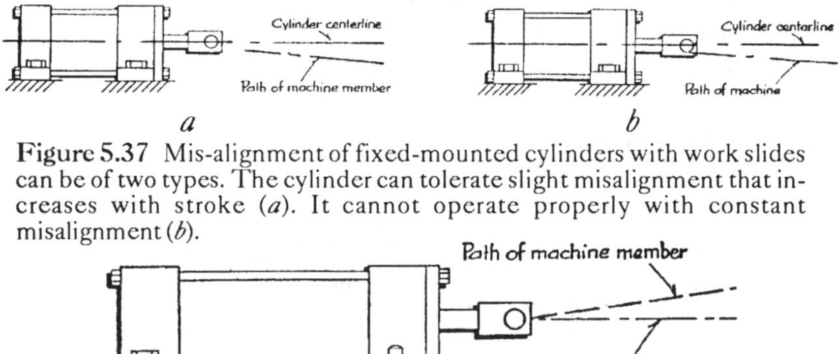

Figure 5.37 Mis-alignment of fixed-mounted cylinders with work slides can be of two types. The cylinder can tolerate slight misalignment that increases with stroke (*a*). It cannot operate properly with constant misalignment (*b*).

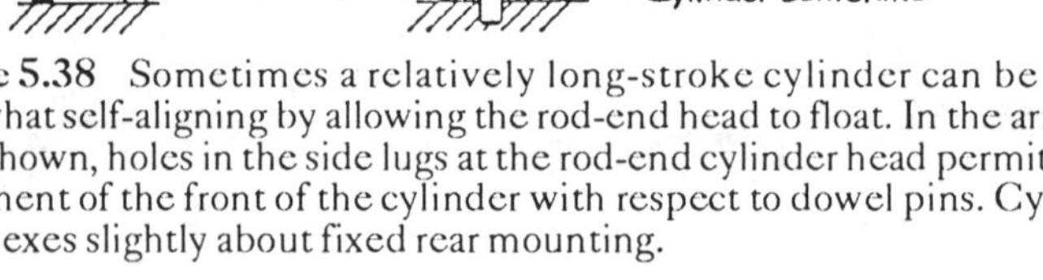

Figure 5.38 Sometimes a relatively long-stroke cylinder can be made somewhat self-aligning by allowing the rod-end head to float. In the arrangement shown, holes in the side lugs at the rod-end cylinder head permit some movement of the front of the cylinder with respect to dowel pins. Cylinder body flexes slightly about fixed rear mounting.

7. In attaching a cylinder rod to a machine member, the rod should not be rotated unnecessarily, as there is always danger of scoring the cylinder body. This is especially so if there is misalignment when the rod is rotated.

8. To facilitate gland maintenance; rod end knuckles, cam surfaces and other projections should not be welded or otherwise fastened permanently to piston rods during installation. Another point is that when seals need replacement the rod bearing probably should also be replaced. Seal wear rates can increase greatly when glands are oversize or worn.
9. Wherever possible, threaded joints should be pulled up tight against shoulders. This will minimize bending at necked-down sections and decrease fatigue stresses caused by reversal or forces in bending.
10. The effects of dirt and other foreign material should be considered at the installation of a cylinder. The installation team should be reminded to ensure no paint or similar material is sprayed on an exposed cylinder rod, and damage to exposed rods should be avoided.

ROD END JOINTS

In specifying a rod end joint, the engineer should carefully consider the thread to be used at the piston rod end. For example, the thread OD should be somewhat smaller than the full diameter of the piston rod to allow replacement of seals without scraping their sealing edge(s) on the threads. Provision of thread relief, rather than run out of threads, provides a stronger rod end and reduces notch sensitivity. The advantages of female threads should be considered (Figure 5.39*a*). Such threads are generally easier to service and if breakage occurs, a higher strength stud can be easily reinserted. In general, fine pitch threads are stronger than coarse threads because of greater total thread root area.

The type of connection of the rod end to the machine member should also be considered. The ideal rod end connection is found where the piston rod is threaded to a well-guided machine member. This arrangement gives the best possible support of the free rod end. Pivoted connections to moving machine members are usually not as critical on alignment as connections which are rigidly made to the members.

Some rod end connections sometimes considered to be self-aligning cannot, in fact, provide such action. An example of this is the slotted arrangement of Figure 5.39*b*. Here, the reduced-diameter section of a rod end engages a slot provided on a machine member that moves in a straight line. Because of necessary clearance at the connection, and nor-

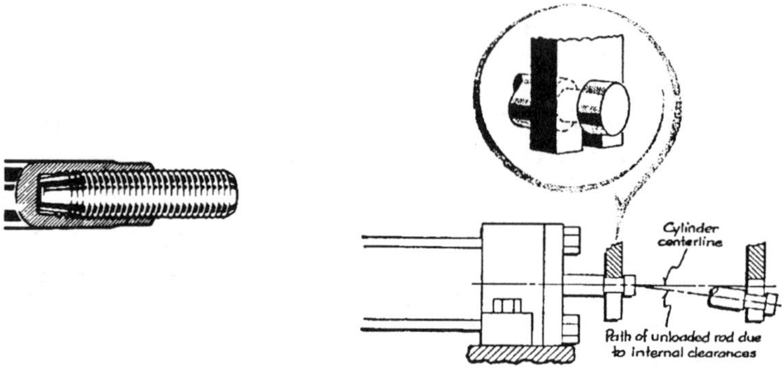

a *b*

Figure 5.39 (*a*) Female rod-end offers versatility and ease of replacement. (*b*) Type of rod end connection shown—generally considered to be self-aligning—can introduce side load on a cylinder. During extension, piston rod-end may follow cylinder centerline because of resistance of machine member and friction at connection. During reversal, however, load is relieved momentarily and rod end can drop to the lower position. On the return stroke, side load—equal to product of coefficient of friction at connection and horizontal pull force—is set-up.

mal clearance between the piston and the cylinder bore, the rod end follows the lower path when no force is being transmitted to the machine member. If the machine member offers sufficient resistance, the rod will follow the upper path during the outstroke of the piston rod. However, during reversal of the piston rod the force at the connection will be relieved momentarily and the rod end will drop to the lower position. During the remainder of the return stroke a side load will be imposed on the piston rod. The magnitude of this side load will be a function of the horizontal pull force required to return the load and the coefficient of friction between the rod end and the slotted member. Scoring of the cylinder body could be one result of such a side load.

PISTON ROD DIAMETER

Selection of the proper piston rod diameter for a fluid power cylinder is based upon several factors including buckling and bearing load conditions, type of rod end joint to be used, and circuit considerations. *Typical points to remember are:* The stroke length of a cylinder may be limited by rod buckling considerations. Here, the rod acts as a long column with one end semi-fixed in the cylinder and the other end fixed or pivoted to the machine member. Where rod buckling can occur, a feasible way to increase the stroke length is to go to a larger diameter piston rod unless the designer is willing to accept operation at a lower design factor.

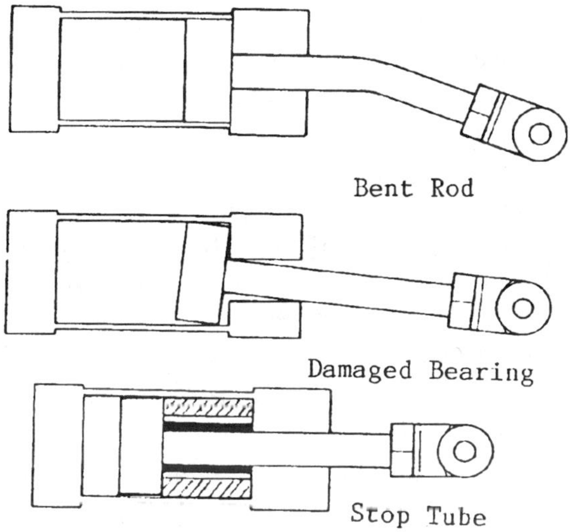

Figure 5.40 A stop tube can limit the stroke of the piston assembly so that mechanical strength is adequate to avoid excessive wear or damaged parts from side loading.

The piston stroke of a specific size rod within the cylinder assembly may be limited by allowable rod end bearing load. These bearing loads must be considered when designing the installation to ensure an acceptable cylinder assembly working life span. Excessive loads may bend the threads at the junction of the piston and piston rod with the rod fully extended. These same excessive loads may lead to early failure of the rod bearings and bore. To limit the stroke to the load-carrying capability of the piston rod, bearings and bore the use of a cylinder with a stop tube will help. The stop tube increases the minimum distance between the rod end bearing and the piston (Figure 5.40).

Vibration and misalignment can require the use of cylinders with large rod diameters. Usually, the bearing area at the rod end gland is increased with larger diameter rods. Sometimes the rod end joint determines the rod diameter selected. For example, if a large shoulder is required at the end of the rod because of high bending forces, only a large diameter rod will suffice. Special rod end joints may also require large diameter rods.

CIRCUIT CONSIDERATIONS

Circuit considerations can affect the choice of rod diameter. The most common example of this is where a differential circuit is used. Usually, the cross-sectional area of the piston rod within a differential cylinder is about one-half the area of the cylinder bore. In some cases, required speed or force ratios for advance and retract strokes call for other cross-sectional area relationships.

CUSHIONING

Pneumatic and hydraulic cylinders are frequently equipped with cushioning devices. One function of a cushion is to provide absorption of the energy of moving masses at the ends of strokes. This includes the masses of the piston and rod, the load being moved, and the fluid medium operating the cylinder. Although it might seem the energy content of the fluid medium is relatively small, it should be remembered that the kinetic energy of a moving mass is proportional to the square of its velocity. Consider for example, a hydraulic cylinder having a 6-inch bore being fed by a 1-inch line. Whatever the speed of the piston, the fluid in the line is moving at a velocity 36 times that in the cylinder. Therefore, the kinetic energy of a given mass of fluid in the line is almost 1,300 times that for the same mass of fluid in the cylinder. This kinetic energy, as well as that of the fluid in the cylinder, the piston and rod, and the load being moved must be absorbed by the cushion in stopping the mechanisms.

From the standpoint of the cylinder manufacturer and the user, cushions also prevent cylinder damage. This is accomplished by the removal of energy over a sufficient length of cushion stroke so that peak forces and pressures will not build-up in sufficient magnitude to damage a cylinder.

A standard cushion will reduce the speed over the last portion of a stroke, but is not designed to provide reliable secondary speed control. Difficulties may be encountered, for example, if very slow primary stroke speeds are involved, since cushions are generally designed to absorb relatively high energy levels in limited time intervals. A cushion may be relatively ineffective or erratic in smoothly and consistently reducing a low primary stroke speed.

Cushioning of pneumatic cylinders often introduces special problems characteristic of the fluid medium used. Bounce is an effect caused by the absorption of the kinetic energy of moving masses without completely exhausting the compressed air which has provided cushioning action. At the moment that motion stops, some air may be stored under elevated pressure and tends to reverse the motion of the piston. This phenomenon may occur several times in rapid succession at the end of a piston stroke. The usual remedy for piston bounce is to adjust the cushion controls so that the cylinder completes its maximum stroke at the instant all of the kinetic energy has been absorbed.

The effect of unswept volume (the air remaining in the cylinder when the cushion has completed its function) should be considered in selecting a cylinder where cushioning is a problem. For example, a cylinder with long cushion engagement may not have the same energy absorbing capacity as a similar cylinder equipped with a short cushion. This occurs where large unswept volumes (with long cushions) reduce the cushioning action. Generally, the greater the unswept volume the lower the compression ratio will be for a given cushion stroke length.

The energy-absorbing capacity of a pneumatic cylinder cushion can often be increased by building up additional back pressure with a meter-out flow control valve; by decreasing the unswept volume of the cushion stroke, and by increasing the cushion stroke length if unswept volume is not increased materially.

Cushioning of hydraulic cylinders introduces problems somewhat different from those encountered with pneumatic cylinders. Bounce effects are virtually non-existent with hydraulic cylinders, since the compressibility of a liquid medium is usually not large enough to cause trouble.

As cushioning begins for the hydraulic cylinder, initial shock may be controlled by suitable entrance tapering on both the male and female cushion parts. Also, the compressibility of a liquid medium is usually sufficient to provide some shock absorption at the beginning of cushioning action.

Machine requirements dictate the type of cushioning action required. Typical cushioning effects which can be obtained with different circuit arrangements are shown in Figure 5.41. Note that the common needle valve method of control may cause a relatively sudden change in piston velocity at the beginning of cushioning action. This could affect machine operation adversely by creating high shock pressures and vibration. For a hydraulic circuit the cushioning action obtainable with a relief valve circuit is more moderate at the initial portion of the cushioning stroke, but increases later. Two special circuits as in Figure 5-42, developed and patented by Parker Hannifin Company, provide the type of action for hydraulic circuits labeled "combination" in Figure 5-41.

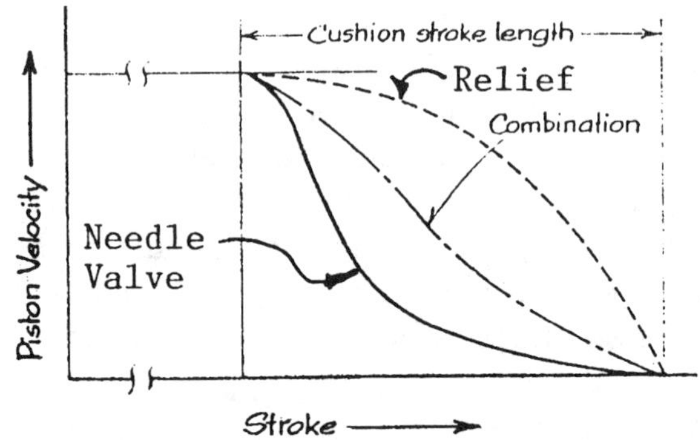

Figure 5.41 Pistons can be decelerated at the ends of their strokes by means of cushioning arrangements. With needle valve, piston velocity may be reduced drastically at start of cushioning (solid line), leading to high shock pressure and vibration. Relief valve method (dashes) gives gentle initial action, but deceleration rate increases during last portion of cushioning stroke. Action labeled "combination" is obtainable with circuits of Figure 5.38.

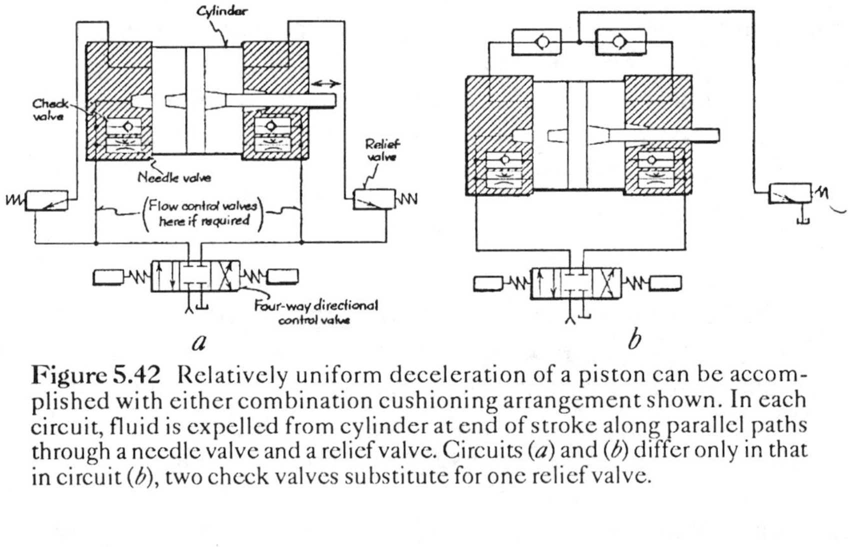

Figure 5.42 Relatively uniform deceleration of a piston can be accomplished with either combination cushioning arrangement shown. In each circuit, fluid is expelled from cylinder at end of stroke along parallel paths through a needle valve and a relief valve. Circuits (*a*) and (*b*) differ only in that in circuit (*b*), two check valves substitute for one relief valve.

ROD SEALS

The functions of a sealing arrangement at the rod end of a fluid power cylinder are to minimize leakage of fluid from the cylinder and to exclude foreign materials from entering the cylinder during the return stroke of the piston rod.

Many different types of seals have evolved to cope with the loss of fluid problem. In the late 1930's, a packing with a "V" shape had been developed to control this leakage. This type of packing, commonly used in hydraulic cylinders, worked relatively well. There was little or no collaring of fluid off the rod, and relatively little dripping.

However, as hydraulic and pneumatic machines became more common in the machine tool industry, it was found that this type of packing was relatively ineffective in excluding foreign materials from the cylinder. Metallic particles, chips, and other foreign materials adhering to the piston rod were too easily drawn past the packing and bearing and into the cylinder. Bearing damage, rod damage, contamination of the fluid, corrosion of parts, and clogging of valves and lines often resulted. Decreased packing life and added maintenance for filter and separator elements occurred.

To cope with the problem of excluding foreign materials, cylinder manufacturers began adding rod wipers to the gland of their components (Figure 5.43*a*). However, this often resulted in the oil film being collared off the rod by the wiper and the formation of oil puddles below the glands of cylinders so equipped.

Wipers have undergone considerable development. Today, a variety of efficient exclusion devices are available. Where materials adhere tightly to reciprocating rods, metal scrapers are often used. One seal design combines a synthetic wiper and metal scraper in a single unit.

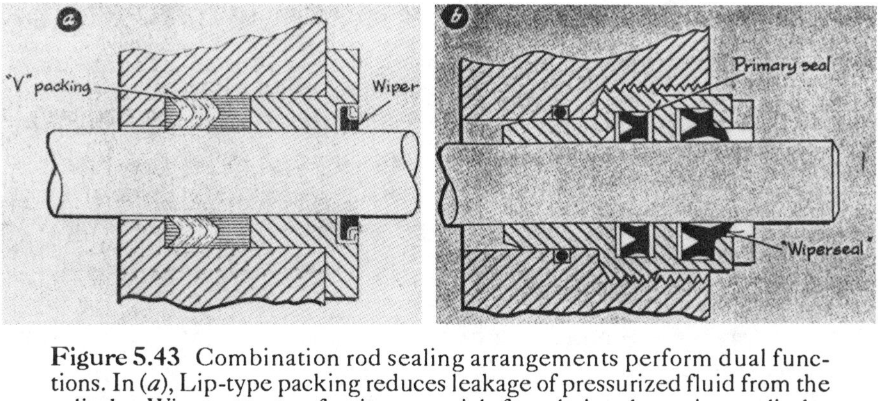

Figure 5.43 Combination rod sealing arrangements perform dual functions. In (*a*), Lip-type packing reduces leakage of pressurized fluid from the cylinder. Wiper prevents foreign materials from being drawn into cylinder during return stroke of piston rod. Arrangement at (*b*) includes two synthetic-type seals in an assembly that performs primary sealing function. Oil getting past primary seal is trapped in cavity formed by inner lip of outer seal and deposited on the rod for return to cylinder during return stroke. Outer lip of outer seal performs wiping function to exclude foreign materials.

A gland sealing arrangement, developed by Parker Hannifin Company is shown in Figure 5.43*b*. This assembly employs a primary inner seal to accomplish the major fluid retention job. Because the primary seal is deformed slightly by the pressure of the hydraulic fluid it is sealing (especially in cases where a meter-out circuit is used), some fluid does get past it. This fluid is sheared off the rod by the inwardly projecting lip of a second, outer seal and stored temporarily in a cavity formed by the lip. Since the outer seal is not subjected to fluid pressure, it is not deformed and therefore does not leak.

An outwardly projecting lip on the outer seal performs the wiping function to prevent the entrance of foreign material into the cylinder. During the return stroke of the piston rod, the oil stored in the cavity of the inwardly projecting lip of the outer seal is deposited on the rod and carried back into the cylinder past the primary seal and rod bearing. The entire sealing assembly, in the form of a removable cartridge gland, is piloted and threaded to the cylinder head. The assembly can be removed without dismantling the cylinder, and either the seals or the entire assembly replaced.

STROKE SPEED CONTROL

Reference was made in Chapter Four that pneumatic speed control does not have the precision characteristics or the versatility of that possible with hydraulic systems. The following information expands upon that discussion.

In most cases the traverse velocity of a cylinder is established by circuit or application considerations. Where the application calls for a definite traverse rate, there is seldom a problem controlling the cylinder within reasonable limits. However, difficulties related to lack of speed control can arise in some cases. For example, high traverse rates that can occur without speed control can introduce problems of stopping the cylinder and load.

The traverse velocity of a hydraulic cylinder may be limited by pump capacity or by means of flow control valves. Speed control by pump capacity is fairly simple if one pump is operating one cylinder. It becomes more difficult if one pump is used to operate several cylinders simultaneously.

The use of flow control valves to limit hydraulic cylinder speeds is particularly important in systems where very large fluid supply capacities are available, as in central supply systems. In a typical system of this type, a large pump and accumulator arrangement serves many cylinders, each of which is controlled by an individual four-way valve. Such a system may operate well when all the cylinders are simultaneously in operation during a normal work cycle. However, where only one cylinder may be actuated at a time, a virtually infinite source of fluid supply becomes available to that cylinder. Piston velocity is then limited only by the length of stroke, since the piston is being constantly accelerated by the fluid pressure available. In such a system, the use of flow control valves is mandatory to prevent damage to the cylinder or machine.

Other hydraulic systems which approach infinite supply characteristics include those where a single pump and accumulator supply a single cylinder (see Figure 5.44). Pump energy, at relief valve pressure is stored in the accumulator during idle cylinder periods. When the control valve is shifted, the combined outputs of the pump and accumulator become available to the cylinder. Here again, use of flow control valves between the directional control valve and the cylinder may be mandatory if the capacity of the pump/accumulator combination is relatively large in comparison to the displacement volume of the cylinder.

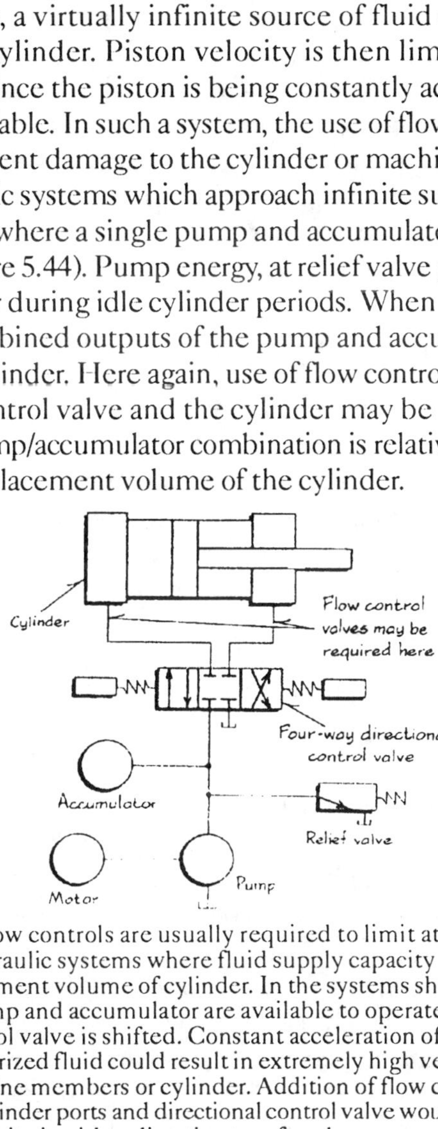

Figure 5.44 Flow controls are usually required to limit attainable piston velocities in hydraulic systems where fluid supply capacity is large with respect to displacement volume of cylinder. In the systems shown, combined capacities of pump and accumulator are available to operate cylinder when directional control valve is shifted. Constant acceleration of piston by large volume of pressurized fluid could result in extremely high velocities causing damage to machine members or cylinder. Addition of flow control valves in lines between cylinder ports and directional control valve would reduce maximum piston velocity in either direction to safe value.

The force buildup characteristics of hydraulic cylinders in circuits which include flow control valves (in meter-in or meter-out arrangements) are such that full force is available on a practically instantaneous basis. This is because the liquid media used in hydraulic systems is relatively incompressible.

Speed controls are important in many applications of pneumatic cylinders. Often, the source of air supply will appear to have a relatively large capacity from the standpoint that the air will expand and continue to accelerate the piston once motion has been started. Some control is possible by means of "flow control" valves, although erratic action can result because of the difference between the coefficients of static and dynamic friction for the cylinder. A degree of speed control sometimes results from serendipity because of the nature of the load or the air supply availability.

An important consideration in the application of a pneumatic cylinder with meter-out "flow controls" is the type of force buildup available at the piston rod for accomplishing useful work. The output force of a pneumatic cylinder stalled or slowed by its load, unlike that for a hydraulic cylinder, increases gradually with respect to time (to allow the effective pressure to build up). See Figure 5.45. The so called "flow control" used in a meter-out pneumatic circuit does not really control the speed. If there is a change in the load because of friction or misalignment - or any other reason - and it results in piston slow down or stoppage - the pressure on the outlet "controlled" side drops until the force is sufficient to break the piston assembly loose - and then it jumps like a shot. If your hand or head is in the way - good bye!

ROTARY FORCE AND MOTION

Pneumatic motors are primarily vane, piston and orbit-type. For some applications the vane type provides the best answer for rotary force and motion (Figure 5.46). The piston design, while more costly can be most efficient for other applications (Figure 5.47). The orbit-type for low-speed, high-torque applications is very small for the power delivered (Figure 5.48*a*). All are similar to their hydraulic counterparts (Figure 5.48*b*, *c*). The typical, rugged, gear-type hydraulic motors are not found in conventional pneumatic applications. The gear structure does not readily lend itself to the normally lower pressures and resiliency of the pneumatic system.

The orbit-type pneumatic motor, like its hydraulic counterpart, is fitted with a generated internal gear set (Figure 5.48*b*). There is one less

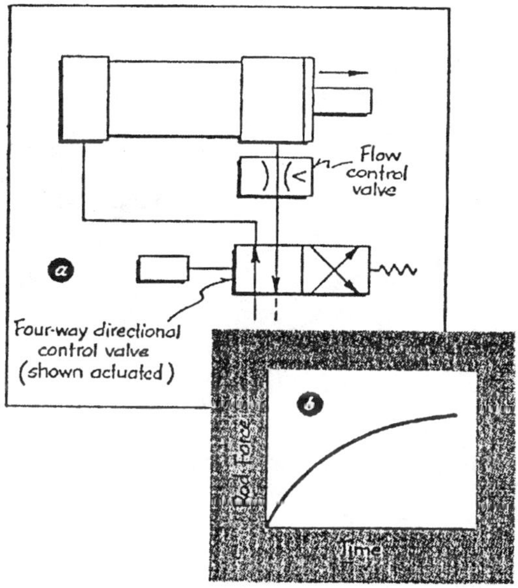

Figure 5.45 Control circuit for pneumatic cylinder includes flow control valve in meter-out arrangement (*a*). With this type of "speed control", build-up of rod force with respect to time is relatively gradual (*b*). Cylinders operated by liquid media respond more rapidly in this type of circuit, and are far less likely to jump uncontrollably because the fluid is less compressible.

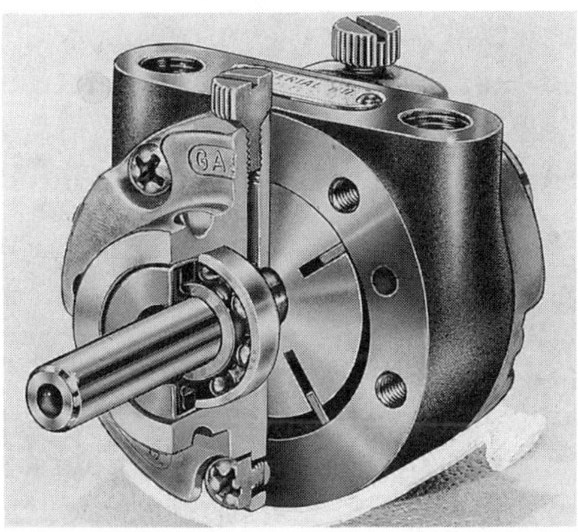

a

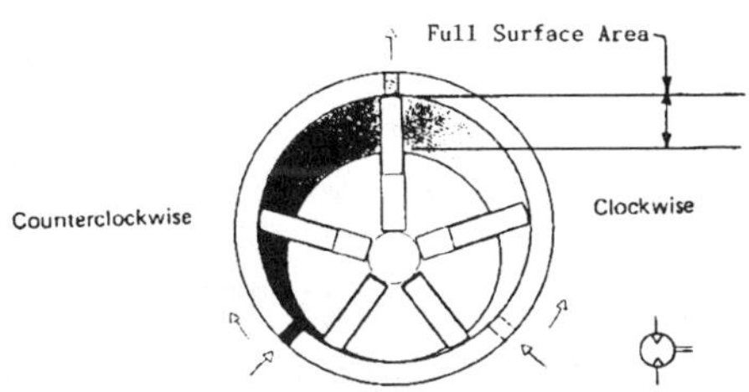

b

Figure 5.46 The vane motor designed for pneumatic service. (*a*) Cut-away view shows the sliding vanes in the rotor following the eccentric path. (*b*) The extending vane offers a greater surface area as the shaft is rotated by the action of the compressed air until it reaches the point of maximum eccentricity. Air is directed to atmosphere as the vane passes the maximum point of eccentricity. Maximum effective surface area is encountered at the maximum point of eccentricity. (Courtesy of Gast Manufacturing Corporation)

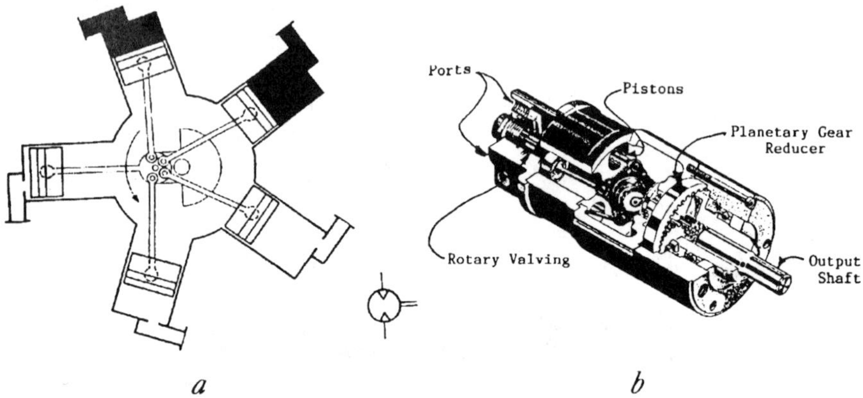

Figure 5.47 The piston motor designed for pneumatic service. (*a*) Air is directed to the appropriate pockets of the radial motor to provide desired rotation by a commutating mechanism. (*b*) An axial piston design can employ a planetary gear drive structure.

tooth on the pinion gear than on the ring gear. The form is generated so that there is always a seal between the tip of each tooth of the pinion and the tooth on the mating ring gear. A commutator valve directs air to the pockets in a pattern that causes the pinion to orbit within the fixed ring gear. A splined shaft serves as a universal joint to deliver the turning force to the output shaft of the motor. The commutating mechanism is also powered by the driven mechanism so that the air is continuously directed to the appropriate pockets to cause the pinion orbit function (Figure 5.48*c*). The mechanism provides an inherent output shaft speed reduction (in this illustration 7:1) and a commensurate increase in torque or turning force.

VERSATILE PNEUMATIC ACTUATORS

Many motion patterns are needed to satisfy the exacting demands of the robots that automate typical manufacturing areas. The actuators of Figure 5.49 illustrate some of the currently available types. The robotic feeding device of Figure 5.50 employs grippers which grasp the product: an up and down motion to grasp the motor part from the supply chute and place it on the armature of the motor. The back and forth motion transports the part from the supply chute to the application point.

The combined rack and pinion assembly of Figure 5.51 for rotary motion (A) and the linear motion by piston (C) provides the two basic motions needed for the transfer function of Figure 5.52. The swing seg-

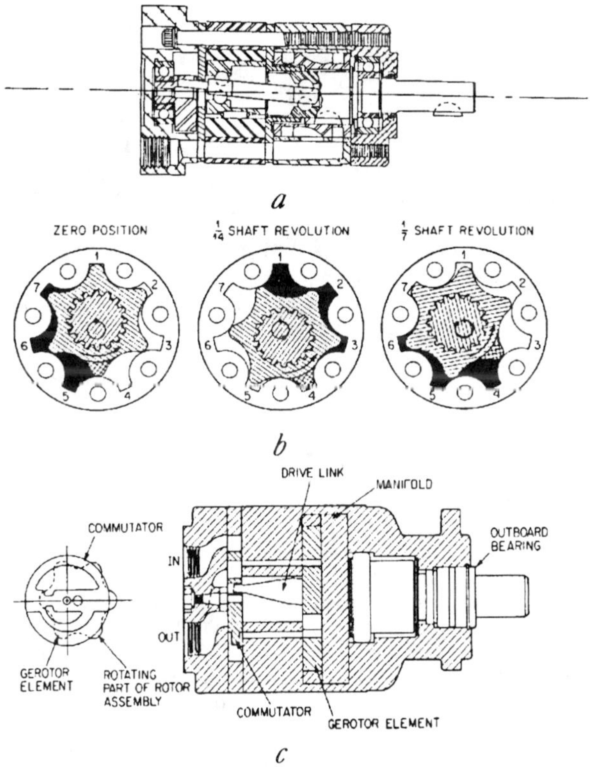

Figure 5.48 The orbit motor designed for pneumatic service. (*a*) Cut-away view of pneumatic orbit motor. (Courtesy of Lehigh Fluid Power). (*b*) The rotor set provides an inherent speed reduction with commensurate increase in torque. Seven orbits of the pinion result in one revolution of the motor shaft. (*c*). The commutator is driven from the output shaft assembly. It directs fluid to and from the appropriate pockets to provide the orbit action and resulting turning force.

ment is fastened to rod (D). Spline (B) transfers the rotary motion to rod (D) while permitting piston (C) to provide the up and down motion. In the same manner the spray nozzle of Figure 5.53 can be moved in an appropriate path both in and out as it rotates to provide the needed spray coverage (the numbers in the small circles show the motion provided).

The applications shown in Figure 5.54 demonstrate a few of the tasks that can be accomplished with the linear and rotary actuator assembly. Figure 5.55 shows the addition of other functions within the assembly processes. Typical multiple motions available to the designer

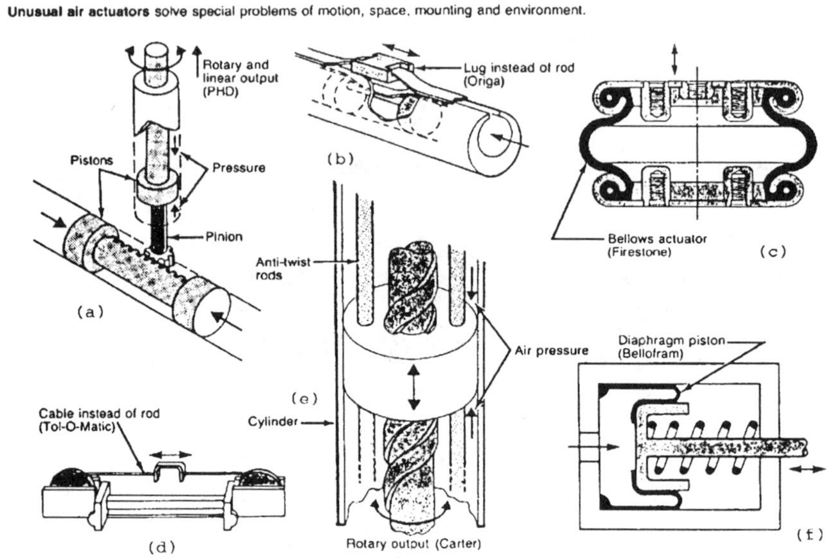

Figure 5.49 Pneumatic actuators. (*a*) Rotary and linear output (Courtesy of PHD, Inc.). (*b*) A lug instead of rod provides motion parallel to the actuator within the same physical length (Courtesy of Origa, Incorporated). (*c*) Bellows actuator provides push with elastomer acting as return because of inherent memory (Courtesy of Firestone Industrial Products Company). (*d*) Cable instead of rod offers space-saving potential (Courtesy of Tol-O-Matic, Incorporated). (*e*) Helix provides rotary motions. (Courtesy of Carter Controls, Incorporated). (*f*) Rolling bellows provides leak-free action without usual rod packing (Courtesy of Bellofram, Incorporated)

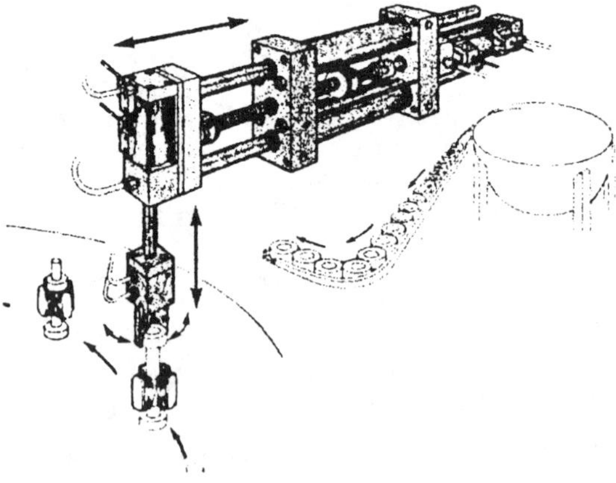

Figure 5.50 Several motions associated with the feeding of materials for electric motor assembly.

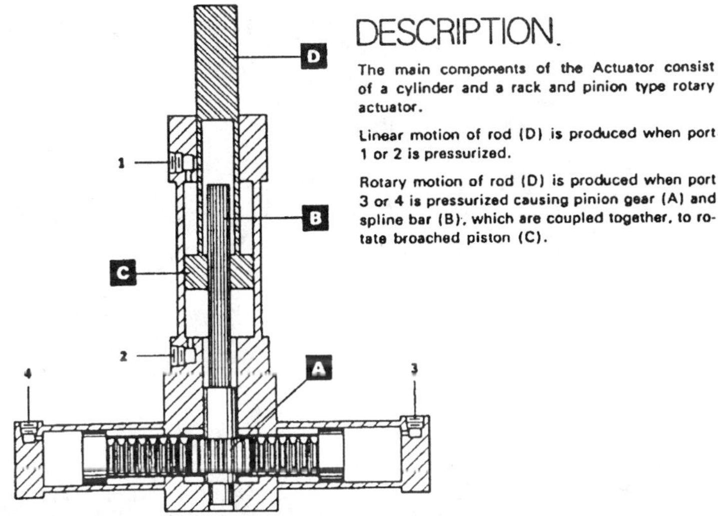

Figure 5.51 Simultaneous rotary and linear force and motion can be accomplished with a combined linear cylinder rack and pinion assembly. (Courtesy of PHD, Inc.)

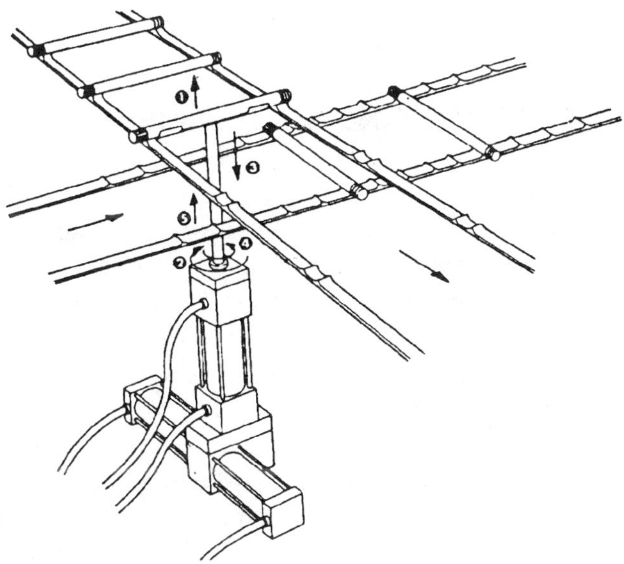

Figure 5.52 Tote box conveyor transfer functions can be accomplished by use of lift and swing assembly of Figure 5.50 (1) lift, (2) swing, (3) lower, (4) swing, (5) lift. (Courtesy of PHD, Inc.)

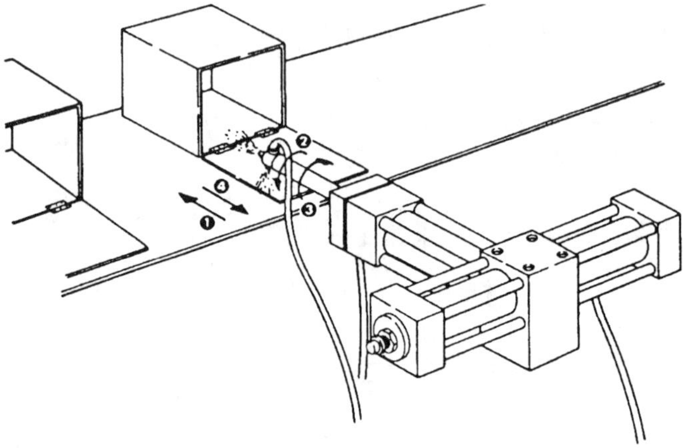

Figure 5.53 Spray nozzle can be positioned to deliver paint in any linear or rotary pattern needed for the specific task. (1) forward, (2) rotate to provide spray pattern, (3) rotate to start position, (4) retract. (Courtesy of PHD, Inc.)

are illustrated in Figure 5.56. The pick-and-place robot of Figure 5.57 has proven to be a good educational device to acquaint students with the capabilities of the robotic structures as well as its obvious value in the work place.

THE CASE OF THE SINGING CYLINDERS!

Figure 58 is an exaggerated view of what can happen when a cylinder is too light for the actual job on which it is applied. An 8, 10, 12 or 14-inch cylinder subjected to excessive pressure above the designed and rated value, coupled with excess torque on the tie rods can permit a change in bore size great enough that the seal on the piston may not stay in contact with the cylinder bore. This is especially true if the circuit is a meter-out where the pressure difference across the piston seal is virtually zero. Seals leak, cylinders scream, move back and forth, etc.

The cure for a "singing cylinder" (depending on the magnitude of the ailment) on a temporary basis may be to put tape (electricians or similar) under the seal ID - fasten the tape to the clean piston surface. This brings the seal diameter up so that it may seal on the O.D. For a quick fix the seal may also be stretched for a larger circumference. Backing-off the tie rod torque may help temporarily. The obvious long term answer - Install a cylinder with correct rating!

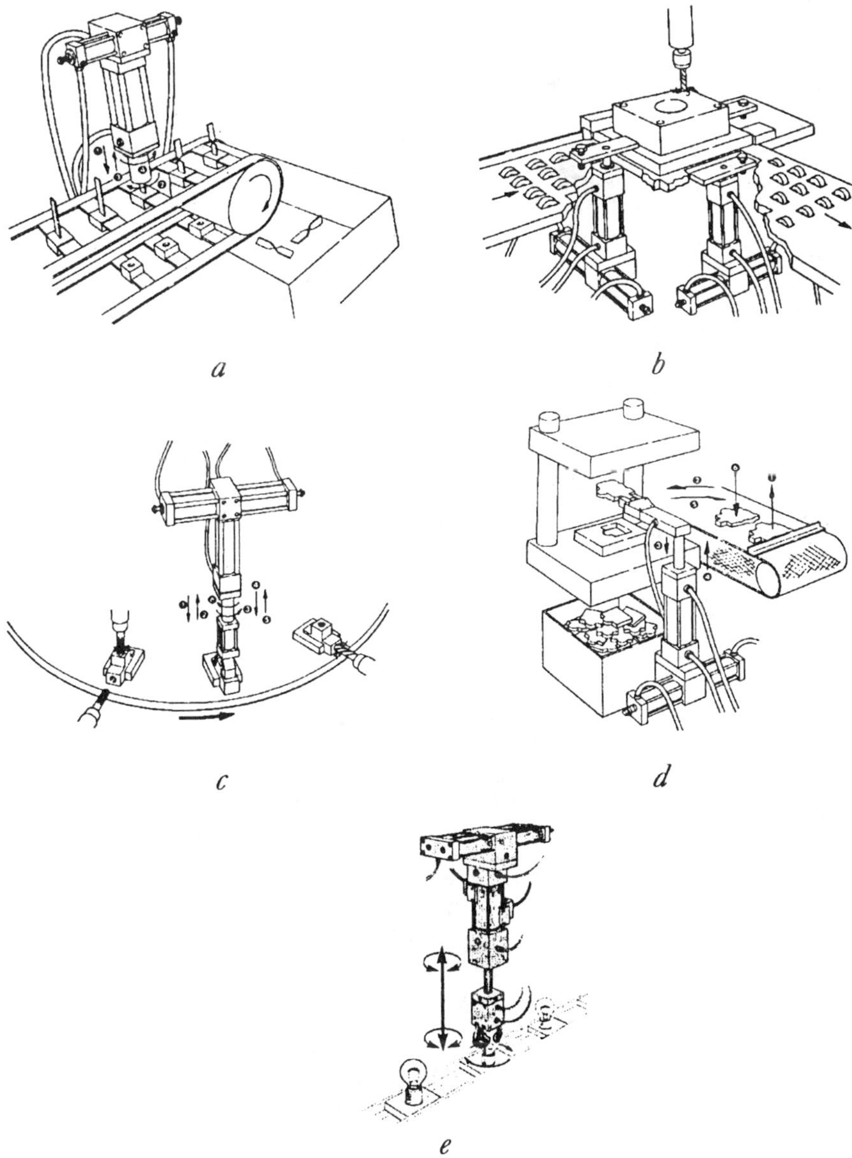

Figure 5.54 Practical applications of linear and rotary motion units. (*a*) {1} Tool approaches task. {2} Part is bent to desired shape. {3} Tool is retracted. {4} Tool is positioned for next cycle. (*b*) Work piece is positioned by conveyor. Clamp member lifts, swings into position and clamps work piece. (*c*) Parts turn around. (*d*). Feeding work piece into trim die. (*e*) The gripper gently clasps a light bulb with a soft grasp and turns it into the socket. (Courtesy of PHD, Inc.)

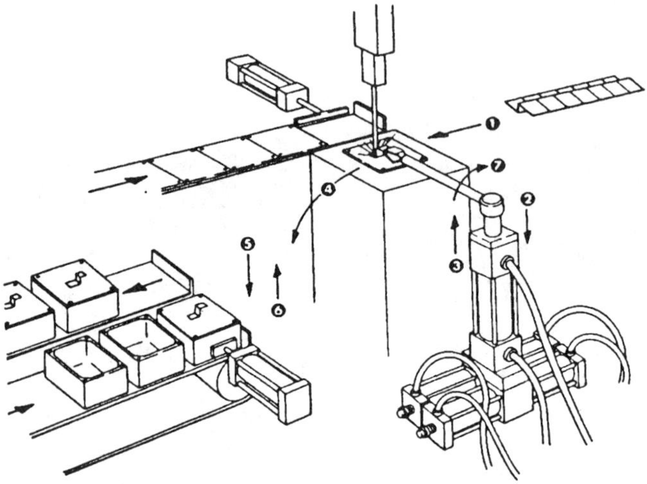

Figure 5.55 Multiposition assembly. (Courtesy of PHD, Inc.)

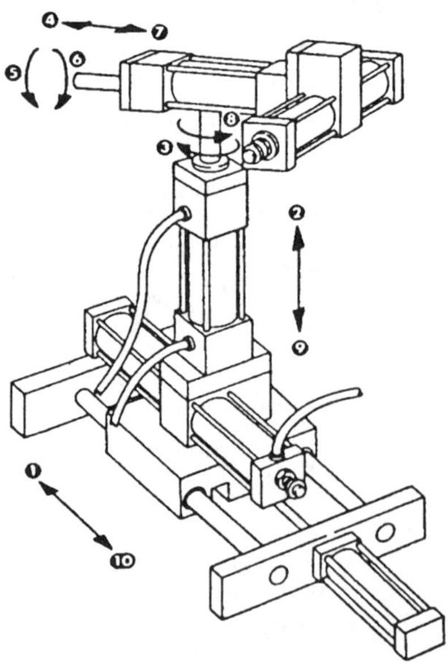

Figure 5.56 Multimotion potentials. (Courtesy of PHD, Inc.)

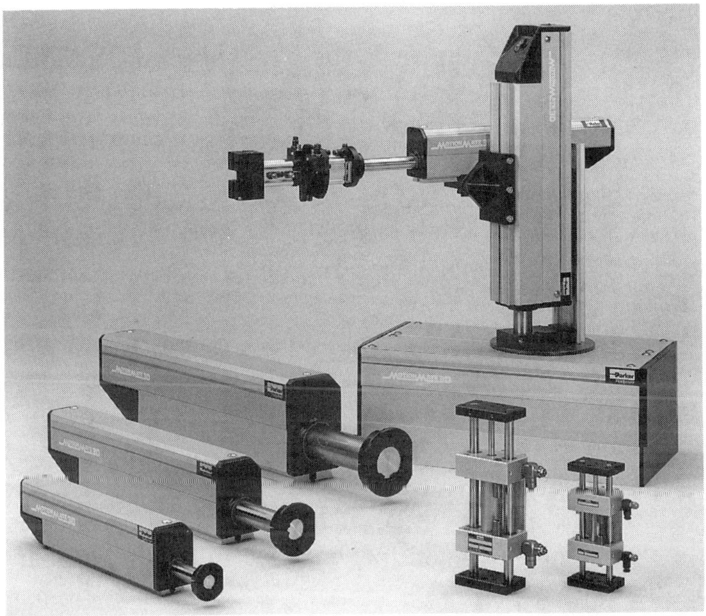

Figure 5.57 Pick-and-place robot. Motion Mate 3D (Courtesy of Parker Hannifin Corporation, 17325 Euclid Avenue, Cleveland, OH 44112)

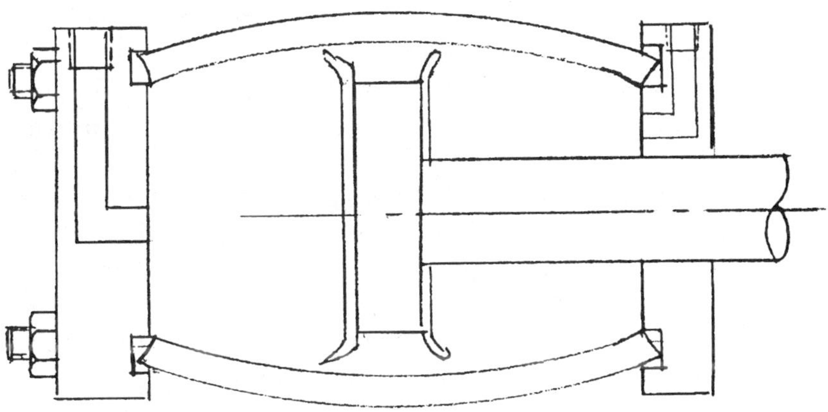

Figure 5.58 An exaggerated view of what can happen when excessive pressure is applied to a large cylinder.

SUMMARY

Because compressed air returns to the atmosphere with minimal contamination, after the power function is completed it can be used for linear, rotary and flotation functions. Pneumatic cylinders, diaphragms and bag devices can provide linear movement. Piston, vane and orbit motors provide rotary force and motion. Flotation devices include the hovercraft-type machines. Air is directed between the surface of the machine to be lifted and moved and another relatively fixed surface acting as the reaction point.

Cylinders must be properly mounted and connected to avoid undesirable stress and wear. Seals and packing must be compatible with the lubricants used in the air compressor and/or the localized lubrication units. Intricate force and motion patterns can be provided with combinations of cylinders, motors, rack and pinion, cables, helixes and screw devices. Air may be used for inflating resilient members such as tires as well as the power transmission function associated with linear and rotary force and motion.

REVIEW QUESTIONS

1 How is compressed air used to float a Hovercraft?
2. What dimensional standards exist for pneumatic cylinders?
3. What are the reasons for using pivoted centerline mountings?
4. Describe a simple method of insuring properly torqued tie rods.
5. What is the purpose of a stop tube?
6. How is foreign matter excluded from a pneumatic cylinder?
7. Name the major problems associated with cushioning pneumatic cylinders.
8. What are the key points to remember when selecting the type of cylinder rod end to be used in a pneumatic installation?
9. What are the advantages of the use of a female rod end when installing a pneumatic cylinder?
10. Where should the cylinder mounting key be located for a major load in tension? For a major load in thrust?

6

COMPRESSING AIR - THE AIR PUMP

COMPRESSSED AIR SERVED THE ANCIENTS

Early hunters developed the blowgun using hollow reeds or bamboo plus other native tube-like structures (Figure 6.1). Using their mouth and lungs, with momentary capacities of about 6,000 cubic inches per minute, they could develop pressures of 1 to 3 psi. Human lungs are a poor compressor for creating a large, useful working force, but the blowgun proved to be an effective hunting aid.

Figure 6.1 The blowgun of the ancients is still used in some remote areas of the world.

Use of bellows have been reported prior to 3,000 B.C. to provide small puffs of air for starting fires (Figure 6.2). These evolved into larger units (but not significantly more sophisticated) used in basic metal smelting about 1,500 B.C.

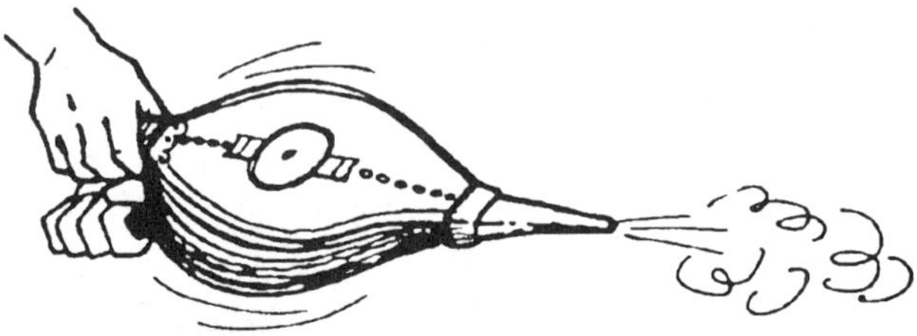

Figure 6.2 The bellows provided greater capacity than human lungs.

THE INDUSTRIAL REVOLUTION PROVIDES REFINEMENTS

Late in the 18th century mechanical compressors achieved the capability of generating pressures as high as 15 psi. In the 1800's there were several attempts to use compressed air as a power transmission agent. The commonly used clay pipe lost so much air in transmission that the value was minimal.

Early in the 19th century, compressors capable of developing 90 psi combined with better piping technology made it possible to use newly developed rock drills to tunnel 8-1/2 miles through Mt. Cenis in the Alps. The original 30 year estimate for manually digging the tunnel was reduced to 14 years with the aid of compressed air and the newly developed pneumatic drills and spades. The tunnel was opened to traffic in 1871.

The role of compressed air in industry has been continually expanding because of the need for flexible, clean power in sophisticated machines which use air under pressure to move cylinders and other types of devices.

The air compressor to many people is the device that provides a source of inflation for automobile tires (Figure 6.3). Such an impression is understandable, as this may be the major exposure to compressed air for most people. The gasoline or electric-powered blower used to clean off the driveway and sidewalk can be thought of as a compressor, but the objective is to move items in a somewhat random pattern. This contrasts to the work done in the industrial sector with captive stored compressed air, or the corner filling station inflating a tire with its small pneumatic supply system. This is not to deny that stored compressed air may also be used to move materials in a random pattern as an additional task for the pneumatic system.

Compressors may be manually operated as shown in Figure 6.4. The basic principle of operation is the same in all reciprocating-type compres-

Figure 6.3 Air compressing can be a chore with a chronic slow leak.

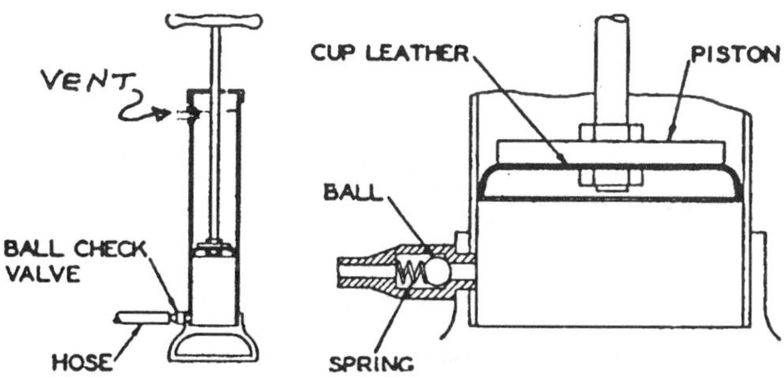

Figure 6.4 A manually-operated air pump.

sors, and the understanding of this simple design can be transferred to other more complex designs of the reciprocating-type.

The hand pump consists of a piston secured by a rod to a handle, a cylinder, a cup seal, and a one-way valve or check. With the handle pulled all the way to the top, air enters a vent hole in the pump cylinder. As the handle is pushed downward, the air now trapped within the cylinder is compressed. As the piston approaches the bottom, the air pressure is sufficient to force open the spring-held check valve, and air enters the system where it is being used. When the piston reaches the bottom of the cylinder and the pressure on each side of the check valve is equalized, the spring closes the check valve. The upward stroke of the piston creates a partial vacuum inside the cylinder. Air flows past the piston seal (acting as a check valve because of its shape) until the inlet port is reached near the top of the stroke.

BASIC COMPRESSOR OPERATION

We can summarize by stating that a compressor is a machine for compressing air or other gases from an initial intake pressure (usually atmospheric pressure) to a higher pressure. Also, a compressor increases pressure exerted by a trapped gas by reducing its volume. The two general classifications of compressors are *positive displacement* and *non-positive displacement.*

Compressors used for pneumatic power are generally *reciprocating piston* or *rotary designs,* which are positive displacement units. Reciprocating compressors have one or more pistons which reciprocate within a cylindrical bore. In a single piston unit, air is compressed from its initial

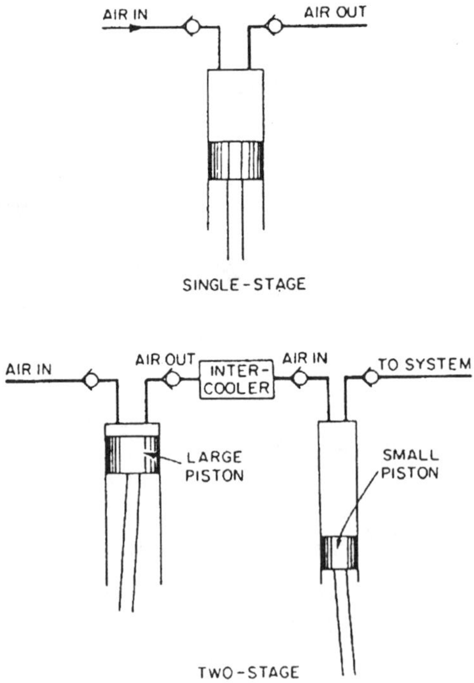

Figure 6.5 Basic compressor operation.

inlet pressure to the desired final pressure in one compressive step. This is known as a single-stage air compressor (See Figure 6.5).

A compressor having two or more compressive steps in which the discharge from each supplies the next in a series is a *multi-stage compressor* (see Figure 6.5). In a single-stage unit the size, stroke and speed of the reciprocating piston determines the volume displacement, generally expressed in cubic feet per minute (cfm). For multi-stage compressors, the piston displacement of the first stage is commonly used as that for the entire machine.

The single-stage compressor of Figure 6.6 illustrates the air cleaner at the input which may be fitted with a replaceable or cleanable cartridge (Figure 6.7). Note the unloader piston which can hold the inlet valve open when the air pressure is at the desired pressure value in the storage tank. The compressor can then continue to reciprocate without compressing air to be delivered into the storage tank. Thus the prime mover need not be stopped. This is important when driving with a heat engine or from a continuous source of power. A simple spring-loaded ball may be used as a relief valve to prevent excessive pressure levels (Figure 6.8).

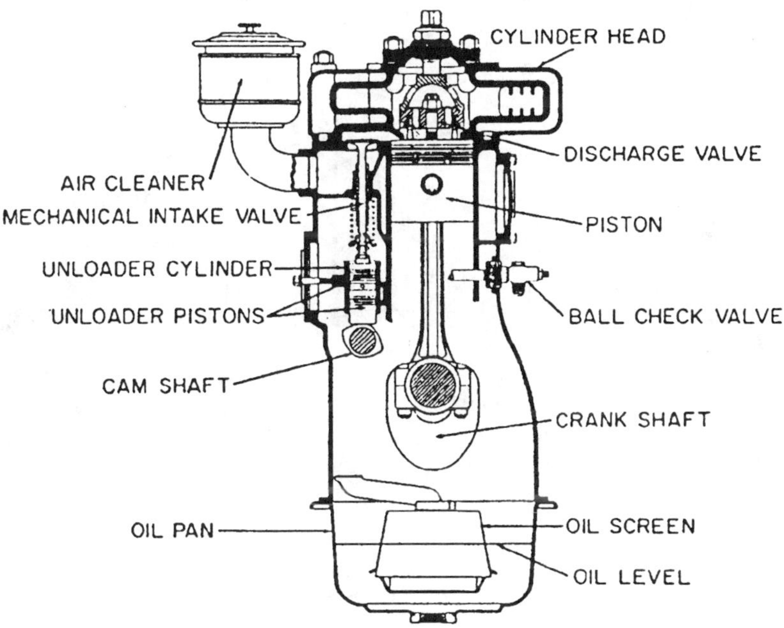

Figure 6.6 Single-stage compressor.

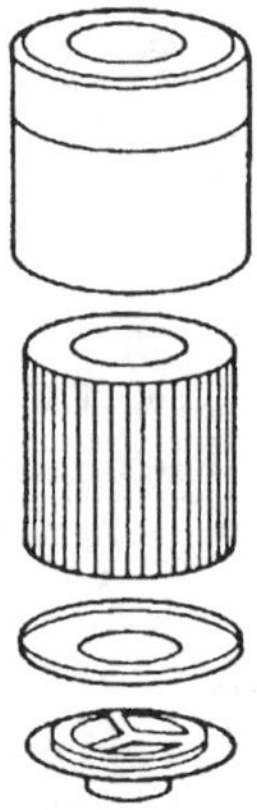

Figure 6.7 Inlet filter with replaceable or cleanable cartridge.

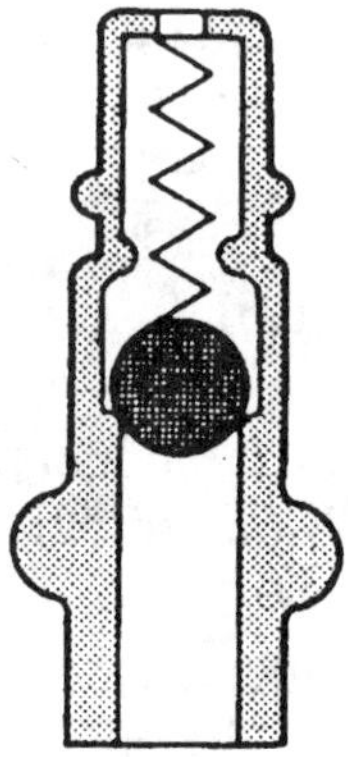

Figure 6.8 A relief valve prevents excessive pressure in the system.

MINIMIZING OIL IN THE OUTLET LINES

Compressing the air is a crucial first step. Cleaning it and drying it will be discussed in a later chapter. Some users with critical applications, e.g. food processing, paint spraying - choose to use compressors that don't need oil for lubrication (Figure 6.9).

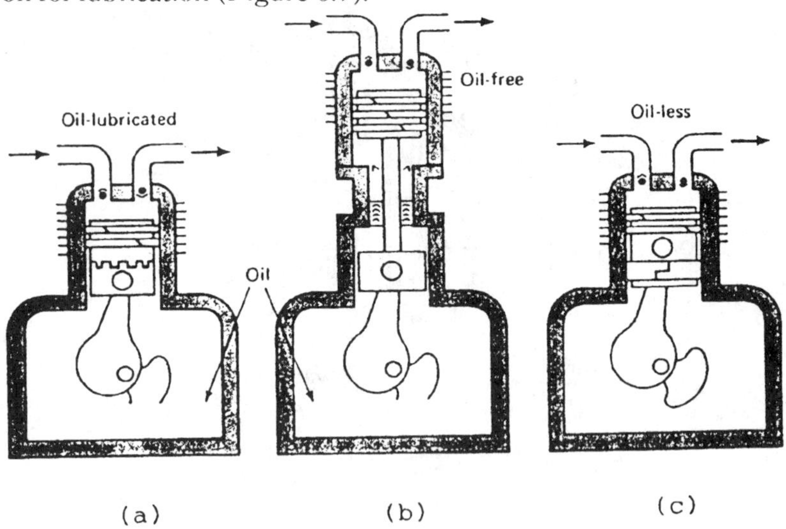

Figure 6.9 Oil-free and oil-less air compressors are distinct from oil-lubricated designs. (*a*) Oil-lubricated. (*b*) Oil-free. (*c*) Oil-less.

Oil, water and dirt in the air wouldn't be too much of an operating problem if all the valves were poppets or other types that tolerate contamination, and if all the working components were built to work on wet, sludgy air. But those days are disappearing. For example, some close clearance spool valves for pneumatic logic become quick victims of drying oil films. Oil in the exhaust air where workers might breath it is a health hazard. If not corrected it may bring on the Environmental Protection Agency inspectors.

Synthetic lubricants within most plant compressors can carry compressor lubricant along with the air being compressed and delivered to the system. It can create compatibility problems with certain seal materials and craze-stressed plastic filter bowls or other parts.

The oil-lubricated compressor (Figure 6.9*a*) will pass some of the lubricant into the receiver and ultimately into the lines. The oil-less compressors, properly applied, practically eliminate the need for oil filters downstream (Figure 6.9*b*). Oil less piston-type compressors frequently have bearing rings of self-lubricating Teflon composites, and the cylinder bores are hard-coated aluminum. Valves may be made of stainless steel, and the head made of treated diecast aluminum. Careful selection of construction materials is needed to prevent moisture corrosion on surfaces unprotected by oil.

The compressor industry makes a distinction between *oil-free* and *oil-less*. "Oil-free" means that oil and oil vapor are kept out of the compression chamber by cross-head seals (Figure 6.9*b*). "Oil-less" means the crankcase is dry or there is no crank case at all (Figure 6.9*c*). The first oil-less (dry crank case and cylinders) piston air compressors are reported to have been built by Bell and Gossett in 1954. The early units had carbon piston rings and permanently-sealed bearings. Later, Teflon-based fluorocarbon devices were substituted for the black carbon rings. The purpose is much the same: to reduce friction without the need of any lubricant.

THE SLIDING VANE COMPRESSOR

Rotary compressors are available in both single-stage and multi-stage configurations. Air is moved via the displacement structure from inlet to outlet. The number of stages and the cooling of the air during the compression cycle (particularly between stages of compression) determines the pressure available at the outlet.

The typical vane design contains a number of rectangular vanes fitted in a slotted rotor (Figure 6.10). The rotor is off center to the housing,

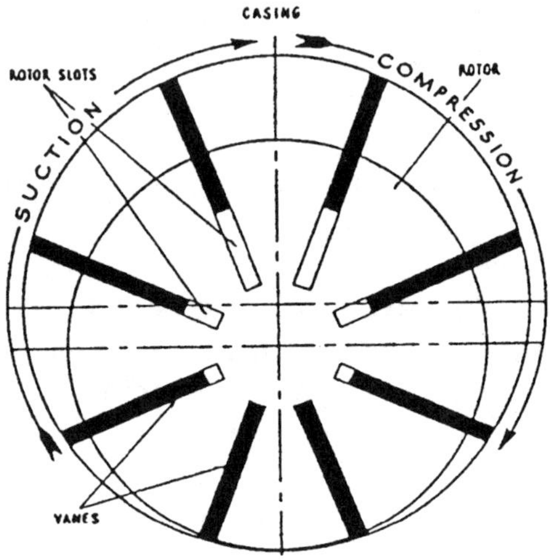

Figure 6.10 Sliding vane compressor.

and as the shaft turns, centrifugal force causes the vanes to move out against the contour of the retaining ring. Air is trapped between the vanes during one-half of one revolution as the space between the rotor and the ring increases. Air is pushed out during the other one-half revolution as the space decreases.

The vanes in a rotary compressor may be made from a fabric cloth impregnated with a phenolic resin. The vanes take on a very high polish during operation and provide excellent wear characteristics. Oil-less models may have carbon vanes or vanes made of "space age" materials based on Teflon, Kevlar and grease-packed bearings.

When air is compressed it gains heat as the molecules of the gas become closer together and bounce off each other faster and faster. Because of the difficulty removing heat in a single-stage unit, there is an economical limit to the pressure level that can be obtained in one compressive step. Generally, this level is 100 to 150 psig. Multi-stage compressors are used when high pressures are required, because better cooling between stages can effectively increase the overall efficiency and significantly reduce the input power requirements.

Screw-type devices (Figure 6.11) are widely used for quiet, efficient, compressed air systems. Some provide very high air flow capacity with low noise level and acceptable efficiency. There are a number of unique designs in addition to the piston, screw and vane-type compressors. The

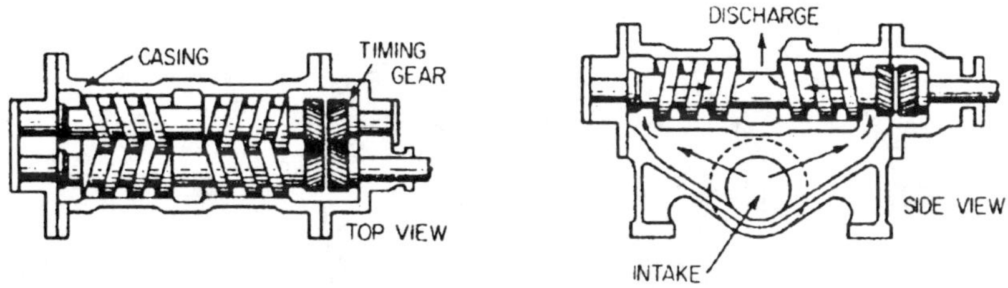

Figure 6.11 A screw-type mechanism with appropriate timing gears provides compressed air with minimum sound level.

two impeller-type compressor of Figure 6.12 is provided with a gear set to keep the impellers in the working positions so that no lubrication is required. These compressors are for relatively low pressure and very high volume, often above 25,000 cubic feet per minute.

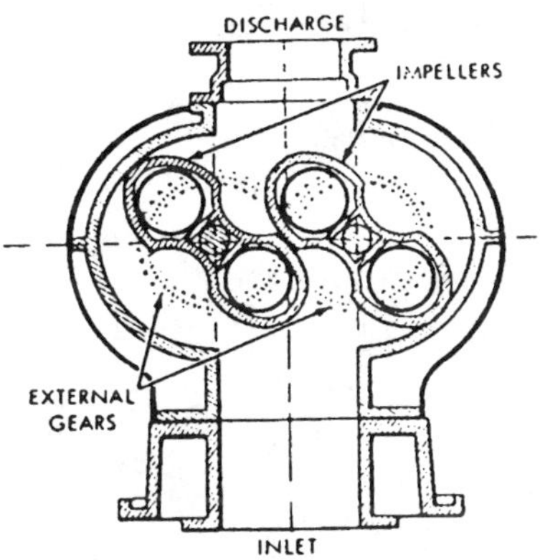

Figure 6.12 The lobe-type compressor moves large volumes of air at relatively low pressures. It also requires timing gears to keep the lobes in the proper position.

The liquid piston-type (Figure 6.13) employs a liquid (often water) as the compressing medium. The blades of the turning rotor form a series of buckets carrying the liquid around the inside of the elliptical casing. Since the liquid follows the contour of the inside of the casing, it will surge back into the buckets at the narrow points of the ellipse. As the liquid surges back, the air in the bucket is compressed and discharged through properly located ports. Compressors of this type develop about 60 psi pressure in one stage. The typical capacity range is up to about 5,000 cubic feet per minute.

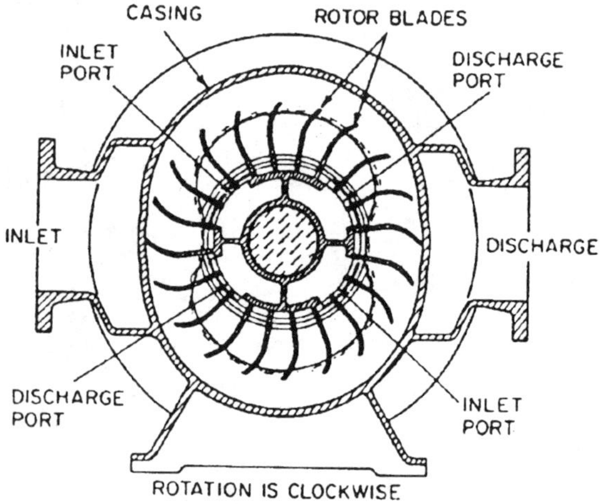

Figure 6.13 A compressor using a liquid in the compression function.

NON-POSITIVE DISPLACEMENT-TYPE COMPRESSORS

A centrifugal compressor accomplishes compression by rotating elements which cause flow in a radial direction. Velocity and pressure is imparted to the air by impellers which may be of the open-, semi-closed or closed-type. Impeller blades may be radial, straight or curved, and they are installed at an angle against the direction of rotation. An example of a four-stage centrifugal compressor is Figure 6.14.

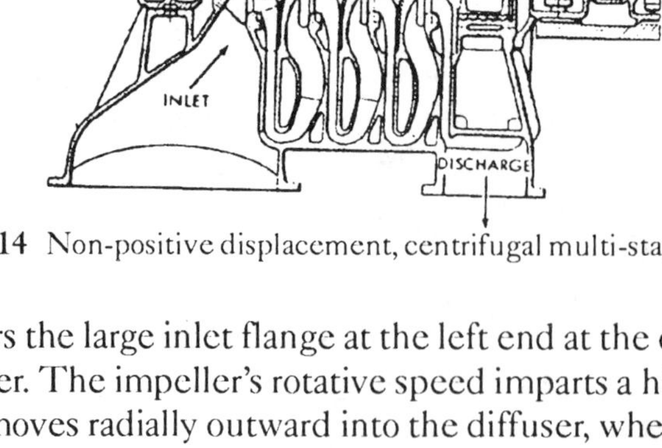

Figure 6.14 Non-positive displacement, centrifugal multi-stage compressor.

Air enters the large inlet flange at the left end at the eye of the first stage impeller. The impeller's rotative speed imparts a high velocity to the air as it moves radially outward into the diffuser, where a portion of the velocity head is converted to a static pressure head. Then the air moves radially inward to the eye of the next impeller and the process is repeated, increasing the air pressure with each compression stage. The axial compressor functions a bit differently. In an axial compressor, compression is accomplished by one or more rows of rotating vanes. A multi-stage axial compressor (Figure 6.15) has two or more rows of vanes which rotate in series on one rotor within a casing, which includes non-rotating vanes for guiding the air to the next stage or row of rotating vanes. Both centrifugal and axial compressors operate at high rotational speed. The former have constant capacity at variable pressures, and the latter deliver practically constant pressure of a wide volume range.

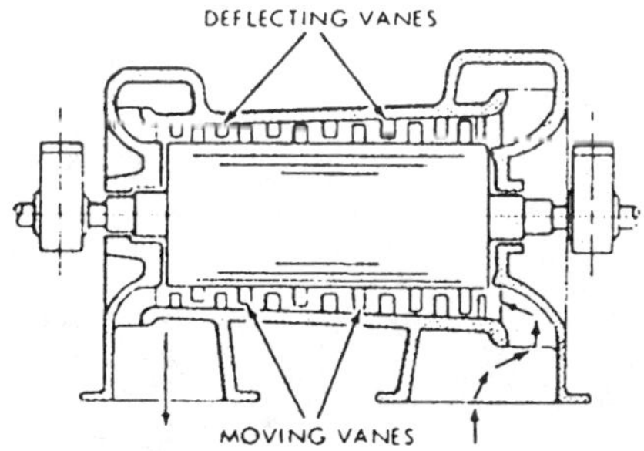

Figure 6.15 Non-positive displacement, multi-stage axial compressor.

UNIT COMPRESSORS

Unit compressors are complete, self-contained units widely used in industrial service, either alone or grouped. The units include the driver (usually an electric motor or heat engine), a V-belt drive or a coupling, a suitably-sized air receiver, valves, fittings, gages and controls for the compressor (Figure 6.16).

PORTABLE COMPRESSORS

Construction activities, demolition functions, tamping, vibrating, and many other tasks can be accomplished with pneumatic tools and devices. The portable air compressor is easily stored, can be towed at highway speeds and be ready to function within minutes of arrival (Figure 6.17).

Many vehicles incorporate an air compressor within their cadre of tools and equipment. It may be mounted permanently on the vehicle or on skids to leave on a job site for a period of time. Service trucks may need a portable compressor (which really becomes a part of the truck) for

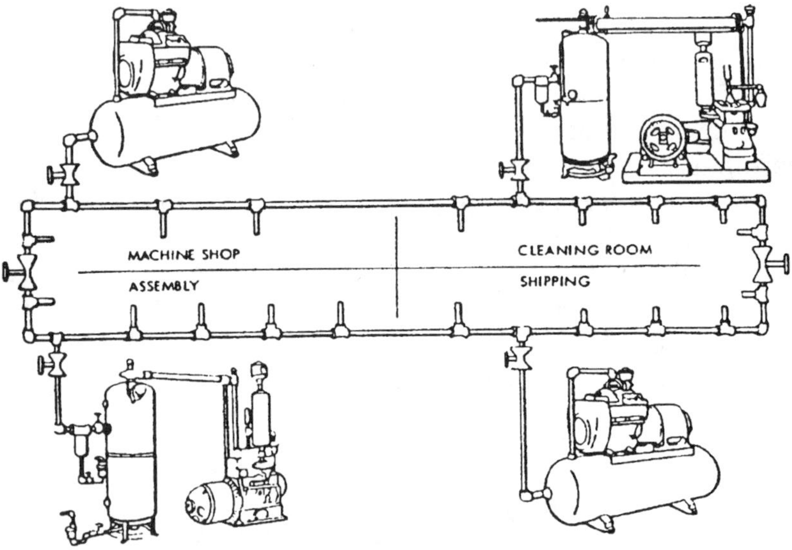

Figure 6.16 Unit compressors connected to shop system.

Figure 6.17 Portable Compressor powering pavement breakers. (Courtesy of Ingersoll-Rand Company)

inflating tires, operating air wrenches, etc. Portable air compressors range from small units that will plug into an automotive cigarette lighter to units requiring multi-axle trailers to support the weight of the diesel or gasoline power source and the compressor. Many tools may be operated from one portable compressor.

Portable compressors can also substitute for or supplement compressed air supplied by stationary compressors. The tapping of main air lines for maintenance work or building construction or alterations can result in lowered air line pressure and the loss of efficiency and production for air-operated production facilities.

SUMMARY

The air compressor provides several functions. If it is to be repaired, replaced or modified the effect on these functions should be carefully assessed, as follows:

1. Is the capacity of the unit affected? Will there be adequate air at an appropriate pressure level to perform the required tasks?
2. Does the system require dry air? Will the replacement unit or the repairs being made affect the quality of the air?
3. Is a change of compressor lubricants being considered? Is a synthetic lubricant used in the replacement compressor? Will this affect the components served by the system being repaired or replaced?
4. Will the sound level be increased by the changes or modifications? Will additional sound level protection be needed to protect workers?
5. Can intake conditions be improved as a result of the changes or modifications? Will the system be more efficient by moving the intake to a cleaner and cooler area?
6. Will additional cooling of the air being compressed increase system efficiency?
7. For good "minimum downtime" performance - a compressed air system installed in a plant or in a long-term job site application should have provisions for quickly connecting an emergency air supply to the system. A capped or plugged "T" should be provided - with shut-off valve and filter to allow for quick connection of an emergency air supply.

REVIEW QUESTIONS

1. Name two general classifications of compressors.
2. What is an intercooler?
3. What precautions should be observed when using synthetic compressor lubricants?
4. What is the difference between "oil-less" and "oil-free" compressors?
5. What is the purpose of multi-staging compressor mechanisms?
6. What is the purpose of an unloader piston and how does it function?
7. What is the typical maximum pressure that can be efficiently attained with a single stage compressor?
8. What is the purpose of the gear set employed in the construction of the compressor of Figure 6.12?
9. What type of compressor would be considered if low operating noise level is required?
10. What type of compressor is designed for high flows at relatively low pressures?

7

EXHAUSTING AIR - THE VACUUM PUMP

VACUUM IS ESSENTIAL THROUGHOUT OUR LIFE SPAN

Each of us at one time may have been considered to be a milking machine. This is common to all young mammals. The farmer, seeking to reduce manual labor on the farm, relies heavily on a vacuum system that makes the mechanical milking machine possible. Those skills learned in infancy carry on with the familiar soda fountain straw used to drink liquids ranging from milk to soda.

THE MECHANICAL VACUUM PUMP LOOKS LIKE AN AIR COMPRESSOR

A vacuum is created in one type of pump by the rotary motion of the off-center rotor, whose four vanes each "bite" off a slice of air at the inlet side; carry it down and around the bottom of the pump, and then compress it so that it exits via the exhaust port (Figure 7.1). Centrifugal force keeps the vanes in contact with the inside wall of the rotary vacuum pump. Notice how the vanes slide out to expose more surface at the bottom of the housing. As the vane continues around the inside of the housing it is pushed into the rotor pocket, exposing less surface as it approaches the top. Each "bite" past the inlet removes more air from the

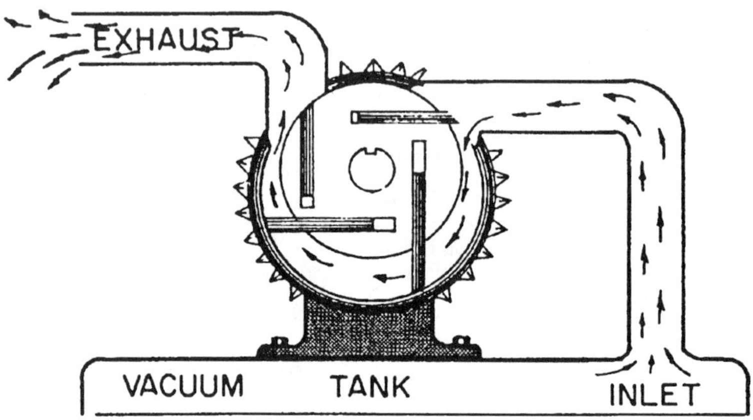

Figure 7.1 The vacuum pump may be considered the inverse of the air compressor.

vacuum working line or reserve tank, creating the desired vacuum required for the milking machine or other vacuum actuated device.

VACUUM IS ESSENTIAL TO THE HOUSEKEEPER

The domestic vacuum cleaner obsoleted the carpet beater and simplified the housekeeper's chores. The vacuum cleaner is not only faster but far more effective than sweeping or other removal processes designed to remove particulate dirt. In each of these instances the basic use of vacuum was in the movement of a liquid or solid. Power was consumed in the action of moving the material.

INDUSTRY USES VACUUM FOR BOTH MATERIAL AND PRODUCT MOVEMENT

The industrial cousin of the domestic vacuum cleaner gets pressed into service to safely remove gasses and solids from the work environment. Such devices may be thought of as an inverted blower. The term *vacuum* is usually applied to any degree of pressure less than atmospheric pressure. If the pressure in an airtight container is 10 psi absolute pressure, it is a partial vacuum. If the pressure is reduced further until it is zero, it would be full vacuum or perfect vacuum. A high vacuum means a very low absolute pressure. Vacuum is commonly measured with a mercury manometer gauge and usually expressed in inches of mercury.

Bourdon tube gauges are also used for measuring vacuum. A compound Bourdon tube gauge reads both vacuum and gauge pressure.

Fluid power in vacuum form is being used for many jobs in industry such as chucking, bottle filling, instrument testing, forming, lifting, feeding, printing machinery, milking machines, and farming operations, plus others.

VACUUM PUMPS ARE TAILORED TO SPECIFIC NEEDS

A four-curved wing-type pump is illustrated in Figure 7.2. A less complex design is shown in the straight vane units. Figure 7.3*a* illustrates a two-vane design. Figure 7.3*b* illustrates a four-vane design. Figure 7.4*a* illustrates a fan-cooled assembly. Figure 7.4*b* illustrates a radiator-cooled assembly. Figure 7.5 illustrates a dual unit to provide vacuum and pressure.

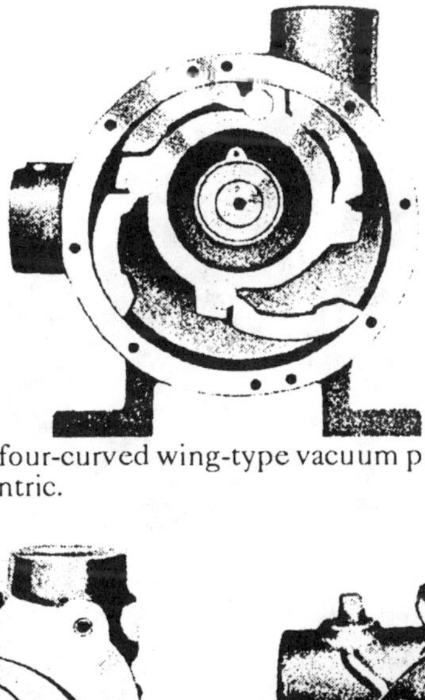

Figure 7.2 The four-curved wing-type vacuum pump provides soft contact of vane to eccentric.

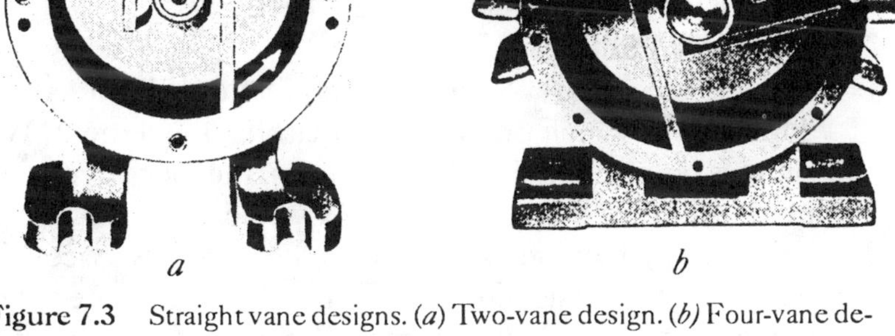

a *b*

Figure 7.3 Straight vane designs. (*a*) Two-vane design. (*b*) Four-vane design.

a b

Figure 7.4 Vacuum pump assemblies. (*a*) Air cooled. (*b*) Radiator cooled.

Figure 7.5 Vacuum and pressure pump assembly.

SOME TYPICAL APPLICATIONS OF VACUUM IN INDUSTRY:

1. VACUUM DOES NOT MOVE MATERIAL BUT AIR VELOCITY DOES

Vacuum creates air velocities in the vacuum cleaner and other applications which provides the material movement. When a sheet is being picked up with vacuum, the pickup force is proportional to the hole area (Table 7.1). The design of the suction head may be modified according to the needed service. Note in Figure 7.6 that a vacuum suction cup may be designed to create the desired area appropriate to the force requirements.

Vacuum inches of mercury	Contact Diameter of Sucker 1 in.	$1\frac{1}{2}$ in.	2 in.	3 in.	4 in.	5 in.
5	1.9	4.4	7.8	17.7	31.4	49
10	3.8	8.7	15.4	34.6	62.0	96
15	5.8	13.1	23.3	52.2	93.0	145
20	7.6	17.3	30.8	69.0	123.0	192
25	9.6	21.8	38.6	87.0	154.0	241

Table 7.1 Holding force created by vacuum on sucker.

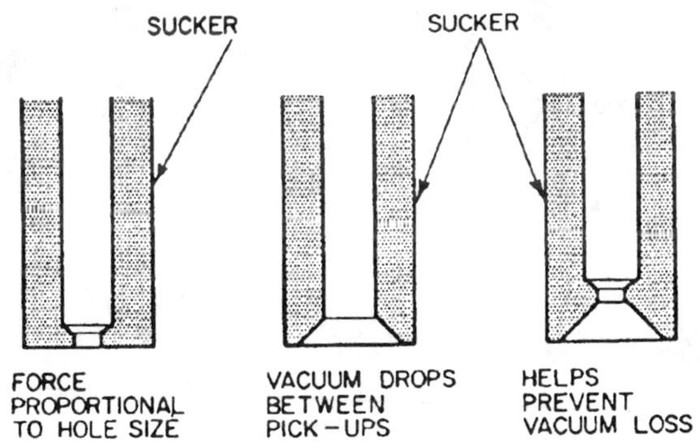

Figure 7.6 Sucker design according to application needs.

2. HOW THE SUCTION CUP WORKS

A suction cup adheres to a surface as the surrounding pressure is higher than the pressure between the suction cup and the surface. The surrounding pressure is practically constant and is equal to atmospheric pressure. To create the lower pressure the suction cup is connected to a vacuum source. The lower the pressure the greater the force pushing on the suction cup (Figure 7.7). Varying designs provide needed holding power for fluffy to heavy loads that can be lifted with appropriate head and vacuum

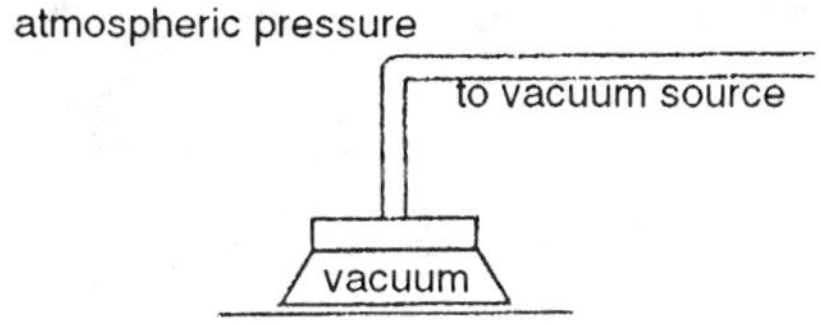

Figure 7.7 The suction cup depends on atmospheric pressure to move the product to the cup as a vacuum source is connected from the pump. This effectively reduces the pressure within the cup so that a significant difference in pressure is created between the outside surface of the product and that portion adjacent to the cup. (Courtesy of Piab USA, Incorporated)

source. Suction cups with cleats are intended for flat, impermeable materials such as steel plate, glass, wood and painted surfaces. They can also be used for certain porous materials such as cartons or fiberboard. When applied vertically the cleats create friction so that the suction cup does not glide over the surface and maintains the full surface area of the cup (Figure 7.8). Very

Figure 7.8 A variety of loads can be handled with the vacuum cup. (1) and (2) Suction cups with cleats are in the first place, intended for flat, impermeable material such as steel plate, glass, wood and painted surfaces. They can also be used for certain porous material such as cartons or fiberboard. When applied vertically the cleats create friction so that the suction cup does not glide over the surface and maintains the full surface area of the cup. (Courtesy of Piab USA, Incorporated)

porous materials such as cotton need large air flows. Small objects such as electronic components can be handled by suction cups with a diameter of a few mm. (Figure 7.9). A suction cup

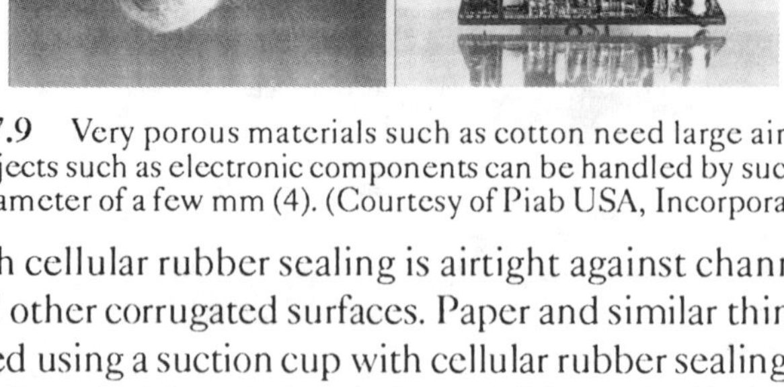

Figure 7.9 Very porous materials such as cotton need large air flow (3). Small objects such as electronic components can be handled by suction cups with a diameter of a few mm (4). (Courtesy of Piab USA, Incorporated)

with cellular rubber sealing is airtight against channelled plate and other corrugated surfaces. Paper and similar thin objects are lifted using a suction cup with cellular rubber sealing and adjustable brace without being deformed (Figure 7.10). A deep suction cup can even lift over perpendicular corners. One type of deep suction cup can easily lift an egg without damage (Figure 7.11). Other types of suction cups with short or long bellows can adapt themselves to most types of curved surfaces (Figure 7.12).

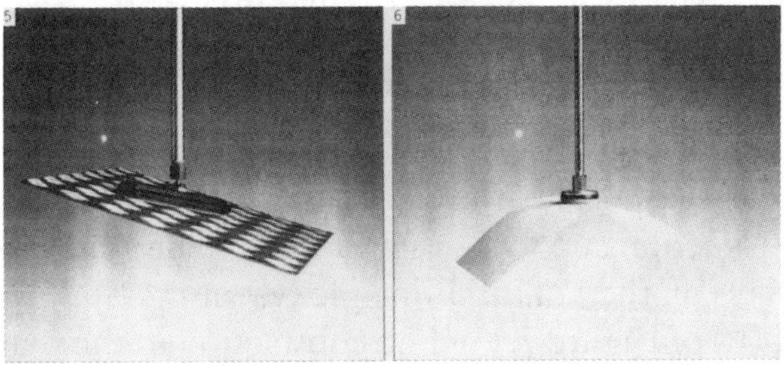

Figure 7.10 A suction cup with cellular rubber sealing is airtight against channelled plate and other corrugated surfaces (5). Paper and similar thin objects are lifted using a suction cup with cellular rubber sealing and adjustable brace without being deformed (6). (Courtesy of Piab USA, Incorporated)

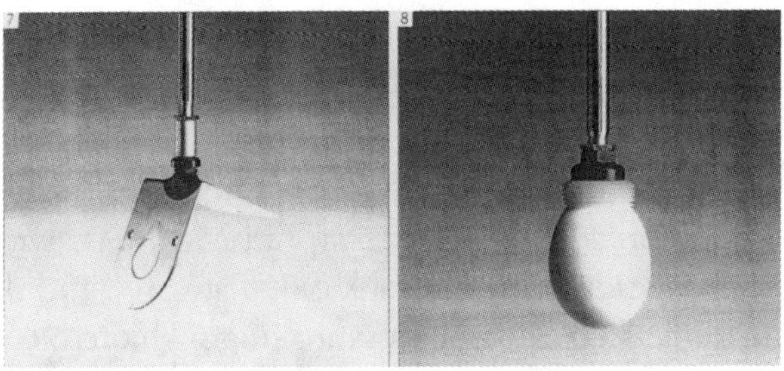

Figure 7.11 A deep suction cup can even lift over perpendicular corners (7). Suction cups with short or long bellows can adapt themselves to most types of fragile surfaces such as an egg (8). (Courtesy of Piab USA, Incorporated)

Figure 7.12 Deep suction cups are used for handling delicate glass and tubular products. (Courtesy of Piab USA, Incorporated)

3. TYPICAL APPLICATIONS OF THE SUCTION CUP

One company identifies a cup needed for opening boxes by a series "B" (bellows) and one for lifting light bulbs as a series "D" (deep) (Figure 7.13). For heavier products they identify the cup as "F" (flat) series and "U" (universal) for lifting corrugated aluminum (Figure 7.14).

4. VACUUM SUCTION NOZZLE

The suction nozzle is used as a transport device in conjunction with a suction cup to move a variety of different components. It operates on the Venturi Principle (vacuum). Supply is applied at inlet (P) of Figure 7.15. The contraction in the diameter of the line increases the velocity of the air flowing toward (R), and a vacuum is generated at suction cup connection (A) (suction effect). In this manner parts can be held and transported. A clean surface yields the best suction effect.

5. VACUUM SUCTION HEAD

The vacuum suction head operates on the same (Venturi) principle. It differs from the vacuum suction nozzle in that it has an integral air chamber (Figure 7.16). The reservoir is filled during the suction process. When the input pressure is switched off, the stored air is released via a quick exhaust valve and flows through the suction cup creating a pressure surge which releases the component from the suction cup.

Air pressure at P directed to the moveable two-diameter tube above the hat-shaped elastomeric cup mechanism forces the hat-shaped member against the hole connecting to the vacuum chamber, and air passes by the lip of the hat-shaped component to fill the reservoir cavity. As pressurized supply air at P is removed, the air in the chamber pushes the two-diameter tube upward, and the captive air enters the vacuum chamber to quickly release the product being held by the suction head.

6. VACUUM FORMING

A sheet of plastic material at an appropriate temperature can be clamped to a female die section. After secure clamping, the air is evacuated from the female half of the die and the atmospheric

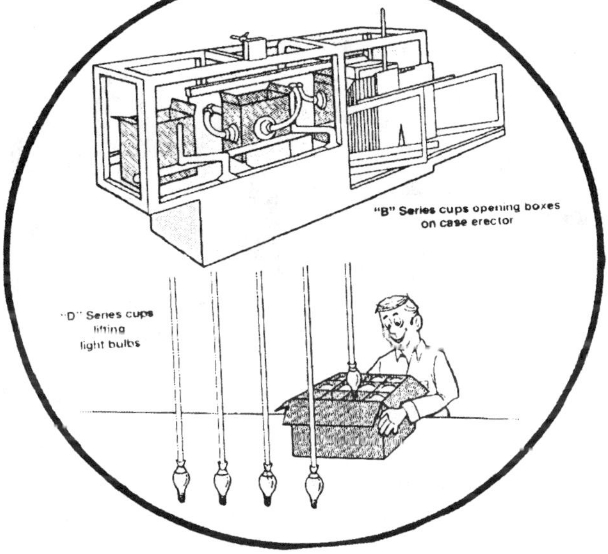

Figure 7.13 Typical application of suction cups for opening boxes and lifting light bulbs. (Courtesy of Piab USA, Incorporated)

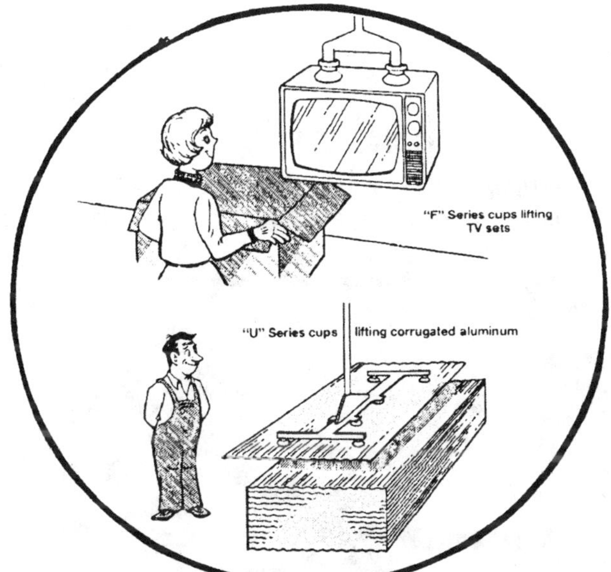

Figure 7.14 Typical application of suction cups for lifting heavy objects and sheet aluminum. (Courtesy of Piab USA, Incorporated)

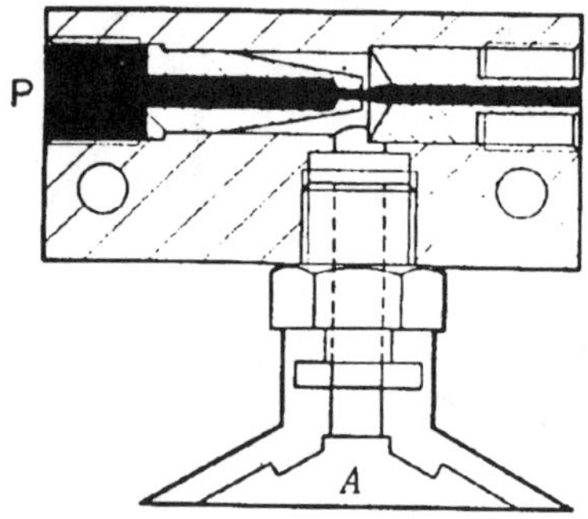

Figure 7.15 Vacuum suction nozzle.

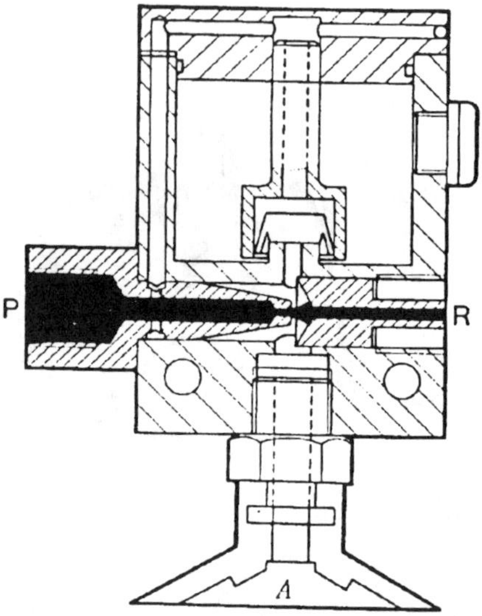

Figure 7.16 Vacuum suction head.

pressure pushes the plastic against the female die surface. Many automotive, household appliance and furniture parts are fabricated in this manner.

THE MULTI-EJECTOR CONCEPT

The vacuum pump of Figure 7.17 is a multi-ejector design that lets the air expand in controlled steps in several chambers. This concept is avail-

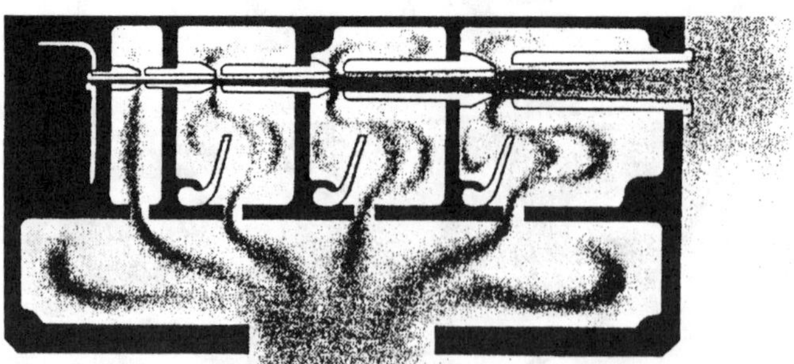

Figure 7.17 Vacuum pump employing a multi-ejector concept. (Courtesy Piab USA, Incorporated)

able in a number of ejector designs with differing characteristics operating in parallel series, so designed as to automatically select the air flow/vacuum optimum. The multi-ejector design has virtually no moving parts which practically eliminates the need for maintenance. Noise level is significantly reduced and no heat is generated because the device is cooled by the expanding compressed air. The size is relatively small, permitting installation close to the point at which the vacuum is needed.

One design features low vacuum for high air flow which is suitable for applications where the system cannot be tightly closed and where only moderate vacuum is required for applications. These include material handling with suction cups, e.g. packaging and printing industries, material handling with some orifices open to atmosphere, vacuum holding, e.g. printed circuit boards and use with robotics with porous surfaces, e.g. corrugated boxes (Figure 7.18).

A second design features a medium vacuum capability suitable for applications where little or no leakage occurs. These include material handling with suction cups, e.g. glass, metal, plastics, wood, etc., robotics with non-porous surfaces, e.g. automobile industry, vacuum-forming, vacuum-holding and placing, e.g. vacuum frame tables and liquid-filling by vacuum displacement (Figure 7.19). An egg is shown being lifted by a vacuum pump with a bellows-style cup (Figure 7.20).

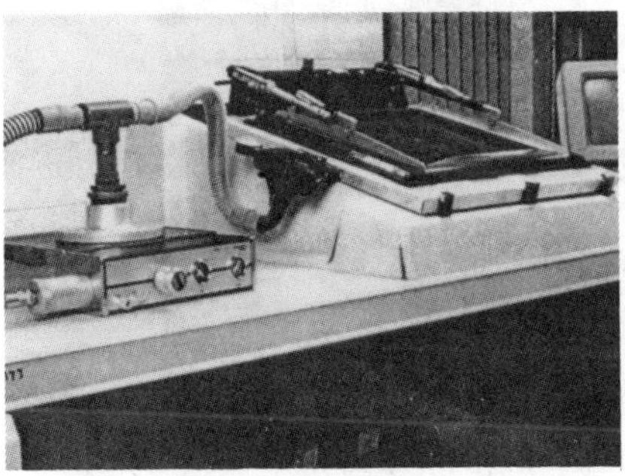

Figure 7.18 Installation of a vacuum pump providing high air flow and moderate vacuum. (Courtesy Piab USA, Incorporated)

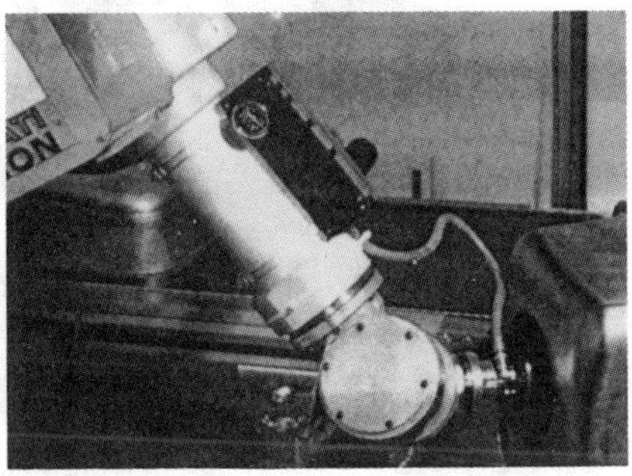

Figure 7.19 Installation of a vacuum pump providing lower air flow and higher vacuum. (Courtesy of Piab USA, Incorporated)

A third series is designed for applications where high vacuums (up to 29.6" Hg.) are required for the pharmaceutical industry, chemi-technical industries, laboratories and industrial finishing (Figure 7.21).

Figure 7.20 Vacuum pump assembly lifting an egg with a bellows-style cup. (Courtesy of Piab USA, Incorporated)

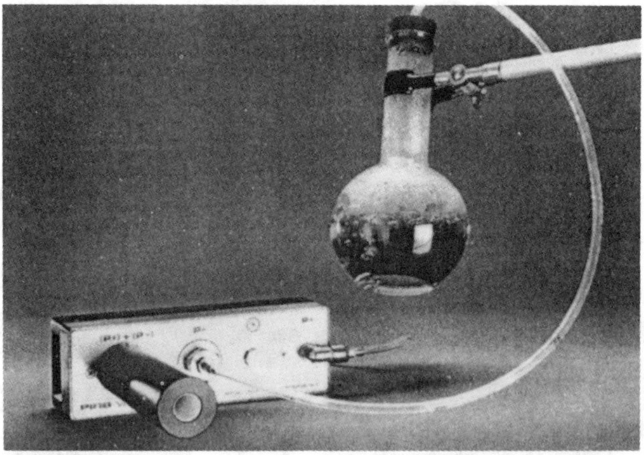

Figure 7.21 Installation of a high vacuum pump for the pharmaceutical and chemi-technical industries, laboratories and industrial finishing. (Courtesy of Piab USA, Incorporated)

FILTERS TO PROTECT THE VACUUM PUMP AND ACCESSORIES

Plastic vacuum filters are available to protect the vacuum pump, switches and valves (Figure 7.22). Metal housing vacuum filters are also available for use in hazardous locations (Figure 7.23). Both units use Polyester felt,

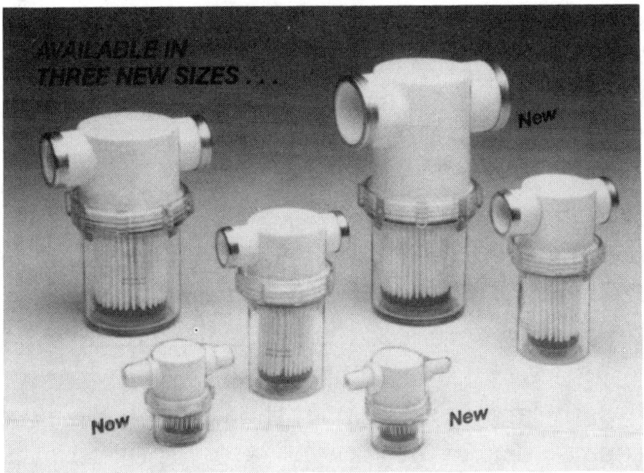

Figure 7.22 Plastic-housed vacuum filters. (Courtesy of Piab USA, Incorporated)

Figure 7.23 Metal-housed vacuum filters. (Courtesy of Piab USA, Incorporated)

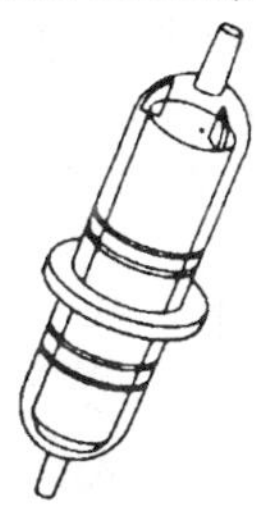

Figure 7.24 In-line vacuum filter for installation in small bore tubing. (Courtesy of Piab USA, Incorporated)

10 micron-pleated paper which is easily removed when servicing. Typically, a high collapse polypropylene core is employed with molded plastic ends in the filter cartridge. Miniature in-line filters are available to fit into small bore tubing (Figure 7.24).

SUMMARY

Vacuum functions are usually an inverse to the compressed air power function. All contaminants generated in the vacuum pump exhaust to atmosphere and the power function is especially clean.

Vacuum pumps can be mechanical or air-powered. Mechanical pumps are usually larger and heavier than the air-powered devices. A supply of compressed air of the appropriate capacity must be supplied for the air-powered devices. Vacuum forming can drastically reduce the cost of forming dies because of the need for only one portion of the die. Vacuum does not move material but air velocity does. The vacuum function creates these air velocities.

Domestic and industrial vacuum systems have materially improved the home and work place. Some homes are piped with a central vacuum system to eliminate the bulky motor and fan assembly that must be moved by the operator from one work area to another. Vacuum systems have kept pace with the robotic devices and automated packaging systems that are now the heart of our industrialized society.

REVIEW QUESTIONS

1. How is vacuum commonly measured?
2. What is the principle of operation of a suction cup?
3. What type of products employ vacuum forming?
4. What is the basic difference between a mechanical vacuum pump and an air powered unit?
5. Vacuum does not move materials. What does move the materials in a vacuum system?
6. What is the basic principle on which the vacuum suction head operates?
7. Can a Bourdon tube gauge be used to measure vacuum?
8. What keeps the vanes in contact with the inside wall of the rotary vacuum pump?
9. Can a Bourdon tube gauge read both pressure and vacuum?
10. Is it true that the term *vacuum* is usually applied to any degree of pressure less than atmospheric pressure?

8

THE STORAGE FUNCTION

THE AIR RECEIVER

The air receiver is the storage vessel to which the compressor or compressors deliver air (Figure 8.1). Because the receiver usually dissipates heat, it may be desirable to position it in an open area where it can radiate as much heat as possible without any adverse affect on surrounding personnel or machinery. The drain should be positioned where it can be readily serviced by maintenance personnel. The receiver is like an accumulator. It stores compressed air under pressure and serves three purposes: 1. It dissipates heat. 2. It coalesces moisture as the air cools (Figure 8.2). 3. It dampens the compressor(s) pulsations and makes the flow smoother and more quiet.

A typical large capacity receiver (Figure 8.3) may be provided with a pressure gauge and thermometer. A pressure relief valve must be provided to comply with OSHA regulations (Occupational Safety and Health Administration). A water drain cock is mandatory. A shut-off valve at the receiver may be useful in servicing the system. A manhole may be provided for clean out purposes.

SIZING A COMPRESSED AIR RECEIVER

The size of a compressed air receiver depends on:

1. Delivery volume of the compressor.
2. Air consumption.

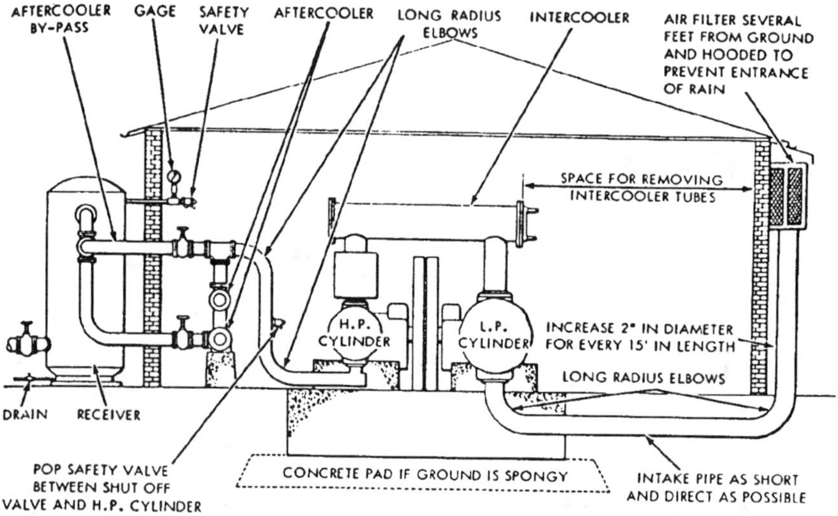

Figure 8.1 The air receiver is the storage vessel to which the compresser delivers pressurized air.

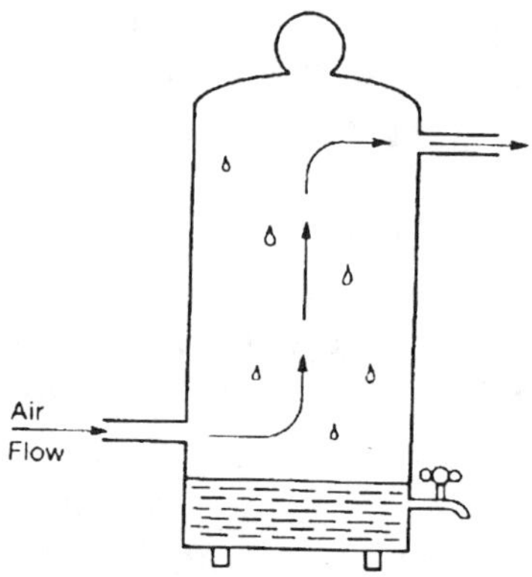

Figure 8.2 The receiver aids in the removal of moisture from the air.

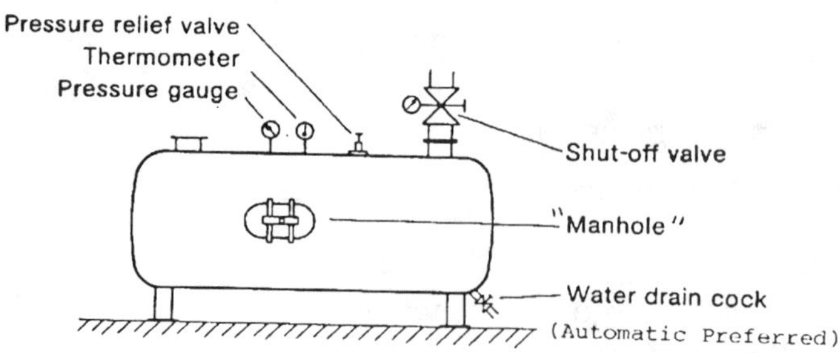

Figure 8.3 Typical compressed air receiver. Volumes are shown in cubic meters and cubic feet.

3. Piping network (whether additional volume may be required).
4. Type of regulation.
5. Permissible pressure difference in the piping network.

RECEIVER DESIGN WITH ON-OFF REGULATION

The receiver size can be determined by using the Figure 8.4 chart.
Example:

Delivery volume	Q = 700cfm = 20 cubic meters/min
Switching cycles/h	z = 20
Pressure difference	– Delta p = 15 psi/1 bar
Receiver size	+ VB = ?

Results: (See Figure 8.4)

Receiver size	VB = 500 cubic feet = 15 cubic meters

DUAL PRESSURE PNEUMATIC SYSTEM

Energy can be saved by supplying only the pressure needed. A secondary feature is lower relative humidity in the second tank (Figure 8.5). These have important advantages in economy and performance. The two-tank system of Figure 8.5 is common in drill rig systems. The first tank is at 175 psi to 250 psi, the second is at 125 psi to 150 psi.

The higher pressure "squeezes" out most of the moisture, which is then automatically removed with an automatic drain valve. The air entering the regulator is saturated with moisture, but as soon as the pressure is reduced the relative humidity drops. Very little condensation occurs from that point on. As an example, reduction of saturated air to 100 psi from 200 psi lowers relative humidity to 53% from 100%.

Freezing still might be a problem with outside rigs, for which there is another trick: an antifreezer. These devices are tanks of alcohol with large porous wicks extending upward into the air stream to allow the air to pick up vapor. Depending on the circumstances, the antifreezer can be placed at the compressor inlet or downstream. *Do not* specify permanent-type antifreeze, however, because the solubility of water vapor in alcohol - a mixture with a much lower freezing temperature than water - is an important characteristic.

AUTOMATIC ELECTRIC DRAIN VALVE

The automatic electric drain valve of Figure 8.6 is designed to automatically drain large amounts of collected water (while minimizing air loss)

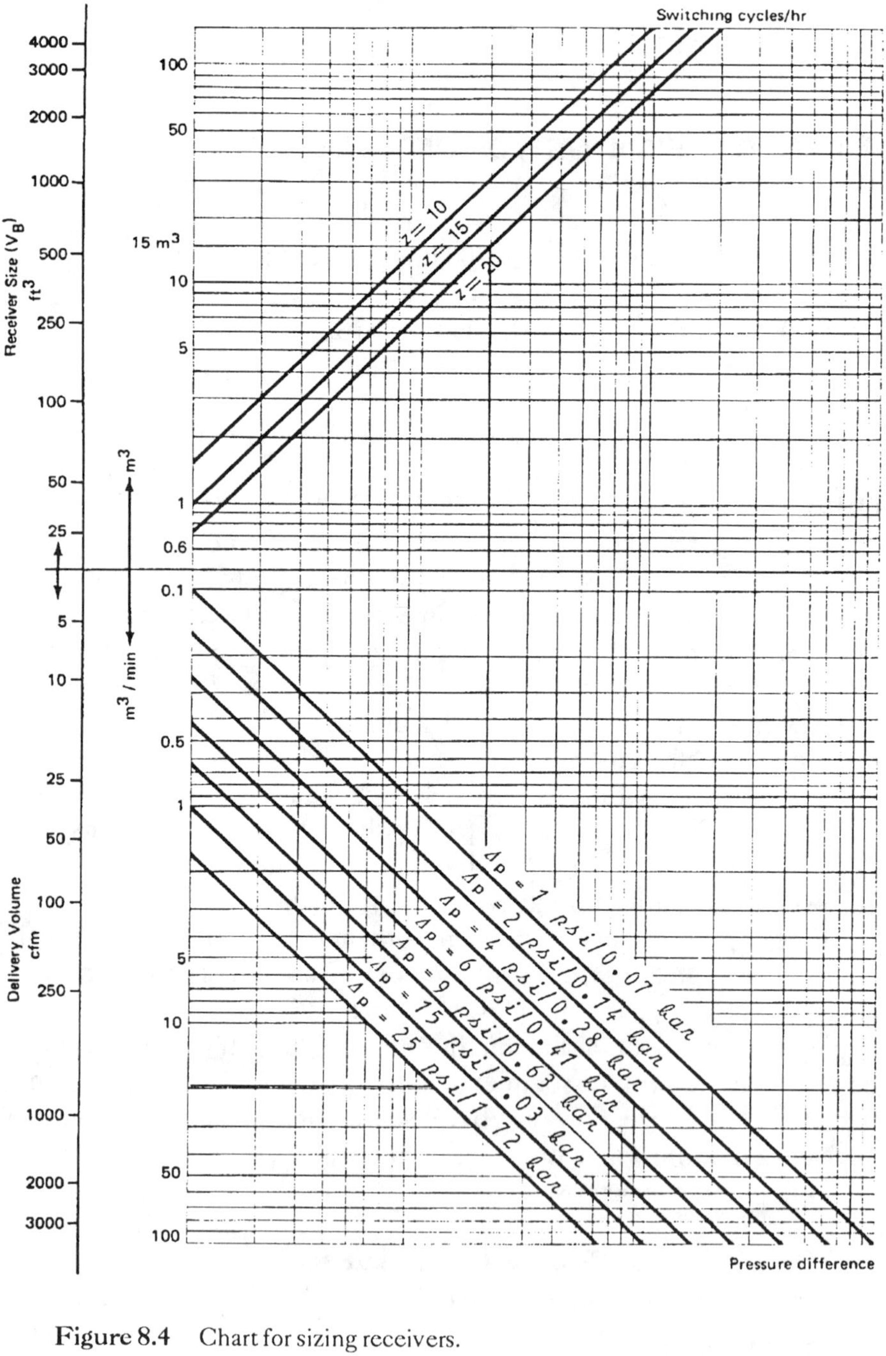

Figure 8.4 Chart for sizing receivers.

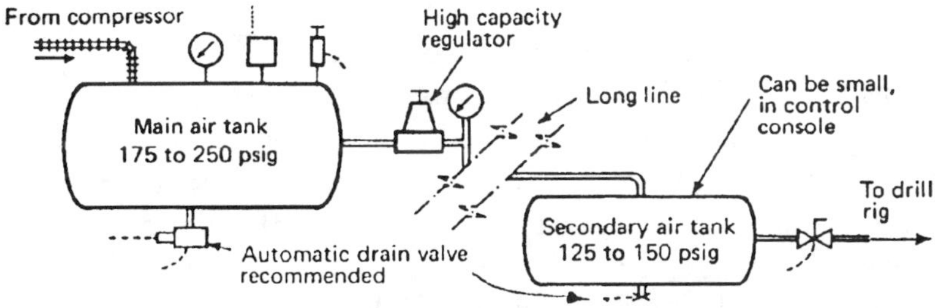

Figure 8.5 Dual pressure pneumatic system saves energy by supplying only the pressure needed. A secondary feature is lower relative humidity in second tank.

Figure 8.6 Automatic electric drain valve. (Courtesy of Wilkerson Corporation)

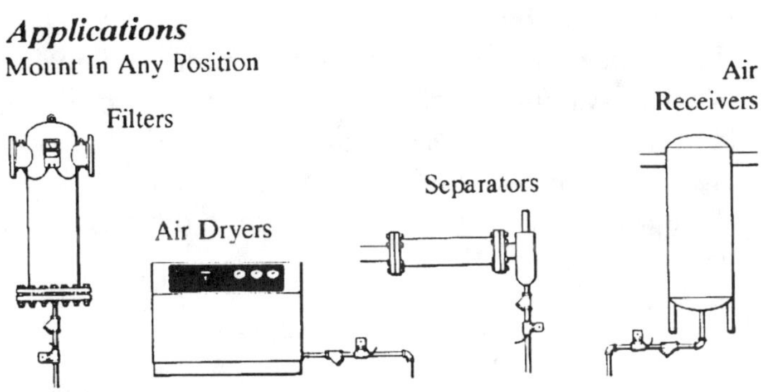

Figure 8.7 Applications of automatic electric drain valve. (Courtesy of Wilkerson Corporation)

from compressed air systems. The valve can be used for draining separators, receivers, refrigerated dryers, filters and drop legs.

The automatic electric drain ensures against damage by liquid carryover. The unit shown is designed to discharge up to 20 GPM at 100 psig. The valve is normally closed. It is designed with a non-clogging full-port 7/16-inch drain orifice to provide reliable draining and minimum routine servicing.

A dual timing feature is incorporated. It can be adjusted from 1 to 45 minutes for cycle time and 1 to 15 seconds valve open time. The valve can be adjusted to an appropriate cycle for the application involved. Typical applications are shown in Figure 8.7. The valve can be mounted in any position convenient to the device being drained.

SUMMARY

An air receiver tank placed in an open area can dissipate heat effectively. Water must be removed as it collects. An automatic water drain cock is preferred. The air receiver is essential to the typical pneumatic system. It dissipates heat, coalesces moisture as the air cools, dampens the compressor(s) pulsations and makes the flow smoother and quieter.

REVIEW QUESTIONS

1. Name three basic functions of the air receiver.
2. What is the purpose of a secondary air tank?
3. Why should an air receiver tank be placed in an open area?
4. Name four factors affecting the sizing of an air receiver.
5. What type of valve is required by OSHA regulations?
6. Is a water drain cock mandatory?
7. What type of device can be used to minimize freezing?
8. What air system components other than the receiving tank can benefit from the use of an automatic electric drain valve?
9. Is an air receiver essential to the typical pneumatic system?
10. What secondary functions occur when the receiver tank dissipates heat?

9

THE DISTRIBUTION SYSTEM

INITIAL SYSTEM DESIGN CAN MINIMIZE FUTURE PROBLEMS

Pneumatic systems in many shops have grown like Topsy. The original installation, quite adequate for a small company shop, may have grown with the addition of new power devices using compressed air. As might be expected, they required greater system capacity as the number of pneumatic devices in service increased.

The simple loop system of Figure 9.1 may have been quite adequate as developed by the designer for the original plant specifications. An expanded system may include multiple branch outlets. A single branch outlet requiring heavy intermittent air use may best be served by an auxiliary receiver to smooth out the usage pattern (Figure 9.2). A drop leg with an automatic air drain will serve to protect the remote station against damage from entrained water gobs.

A LOOK TO THE FUTURE

Supplementary compressors may be added periodically and temporary portable compressors may be connected to the air system for emergency work loads. Thus the relatively inexpensive provisions made for such additions as tees and shut-off valves when the original system is installed can later save significant time and money at a modest first cost.

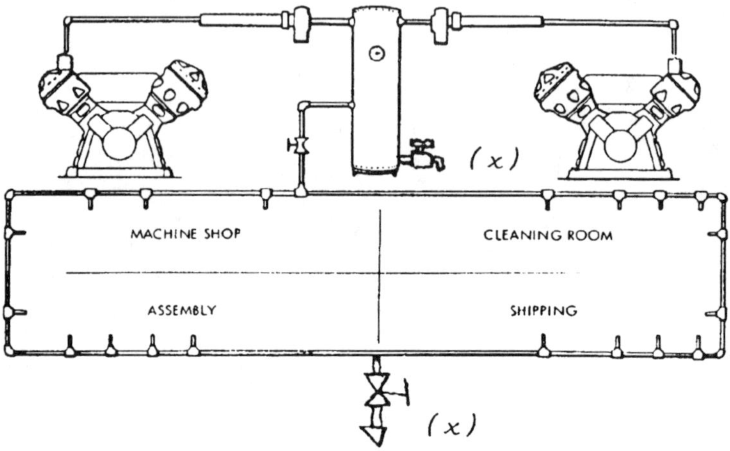

Figure 9.1 Loop pneumatic distribution system for a small shop. A drain - preferably automatic - should be installed at the bottom of the receiver. All take-off connections should be from the top of the line. Air intake should be in a cool dry area if possible. Note the volume of the pipe loop serves as a reservoir and improves the system dynamic performance. Use an automatic drain (x) to the bottom of the line in each sub circuit and at the furthest point from the compressor. (Courtesy Parker Hannifin Corporation)

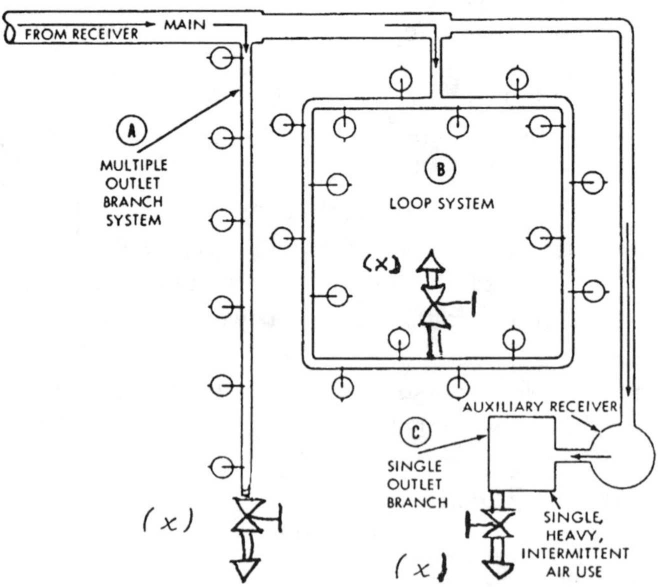

Figure 9.2 An expanding system may benefit from an auxiliary receiver adjacent to a heavy single intermittent air use. Install an automatic drain (x) at far points from the compressor.

SELECTING PIPE DIMENSIONS

Using an outmoded rule of thumb or determining choice of pipe size by availability of pipe in the factory stockroom may decrease the efficiency of a pneumatic system. The following criteria should be considered as the system is planned:

1. Volume of air required for the units to be supplied.
2. Pressure required at the point of usage.
3. Pipe length (distance from supply).
4. Restrictions in the line (e.g. - water separators, filters, regulators, valves, bends, changes in line size).
5. Allowable pressure drop.
6. Water removal and filtration.

A nomogram (Figure 9-3) is used to quickly and easily determine the piping diameter. The nomogram takes into consideration the pipe length, intake volume, inside pipe diameter and shows pressure loss. A straight line from the pipe length column (A) through the intake volume (B) to axis 1 (C) establishes a basic relationship of intake volume and pipe length. The second line from axis 1 (C) through (D), the inside pipe diameter continues through (E), the pressure value, and on to axis 2 (F). From the psi value (E) a line is established through axis 2 (F) to the pressure loss information (G). From this information the frictional losses of the air passing through the pipe can be estimated for a system. At best - all the numbers are approximations useful in the normal working pressure range of 30 to 120 psi. Decisions can be made as to the need or advisability of increasing the pipe size.

Pipe line resistance created by elbows, tees, valves, etc. can be related to equivalent lengths of pipe. The nomogram of Figure 9.4 provides a convenient way of obtaining relative values.

PIPING AND INSTALLATION OF AUXILIARIES

Main air lines must be sufficiently large to prevent excessive pressure loss at the maximum required flow rates. Throughout the air distribution system, one of the most important factors to recognize is the extent to which air pressure drops as the rate of air flow increases.

Compressed air moving through the piping system loses water by precipitation as the air cools. This water should be trapped or otherwise removed from the system to prevent damage to tools and equipment. To help accomplish this, the main lines should slope downward toward the

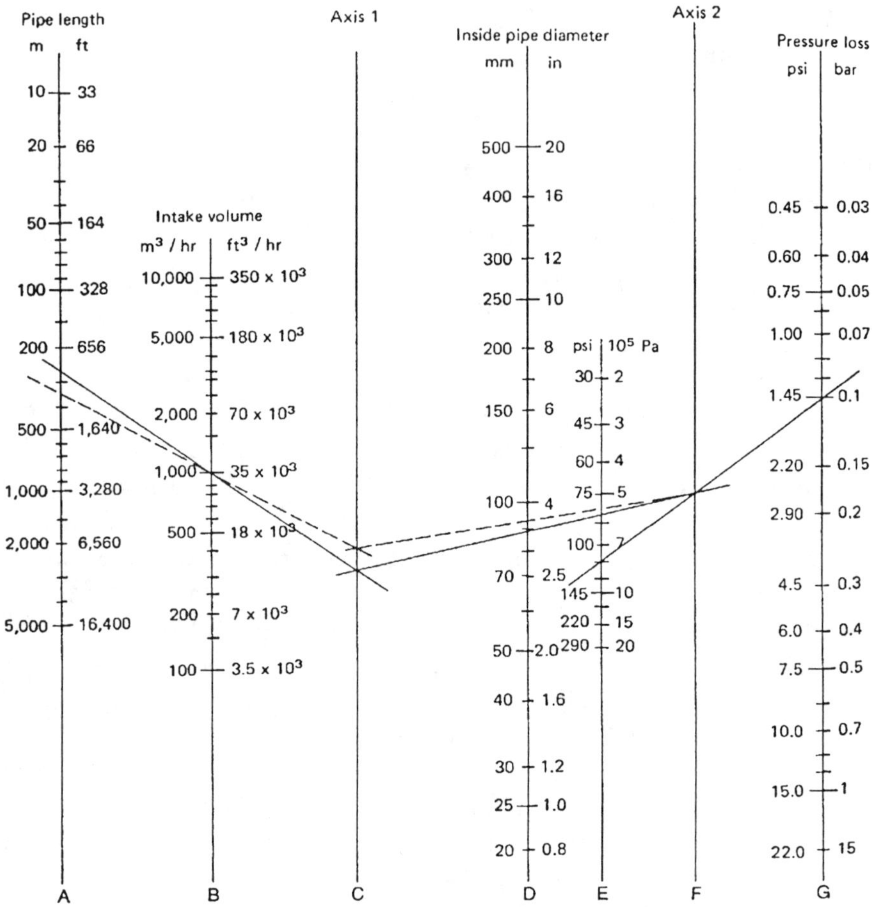

Figure 9.3 Nomogram. Pipe length (Column A) referenced to intake volume (Column B) provides a reference point on axis 1 (Column C). From the reference point on axis 1 a line through the inside pipe diameter measurement (Column D) establishes a reference point at axis 2 (column F). A line from the established pressure value (Column E) through the reference point on axis 2 to the pressure loss (Column G) provides the information relative to line losses.

air receiver or to a low point in the piping so that condensate will be carried to traps, separators or water legs located at various intervals within the system as shown in Figure 9.5. Down- or drop-lines should be taken off the top of the main lines and extended below the feeder lines. Automatic traps, if used, eliminate the need for manual draining of water legs or traps. If automatic traps are installed to discharge into sewers, they must be kept in good working order to prevent loss of compressed air. In

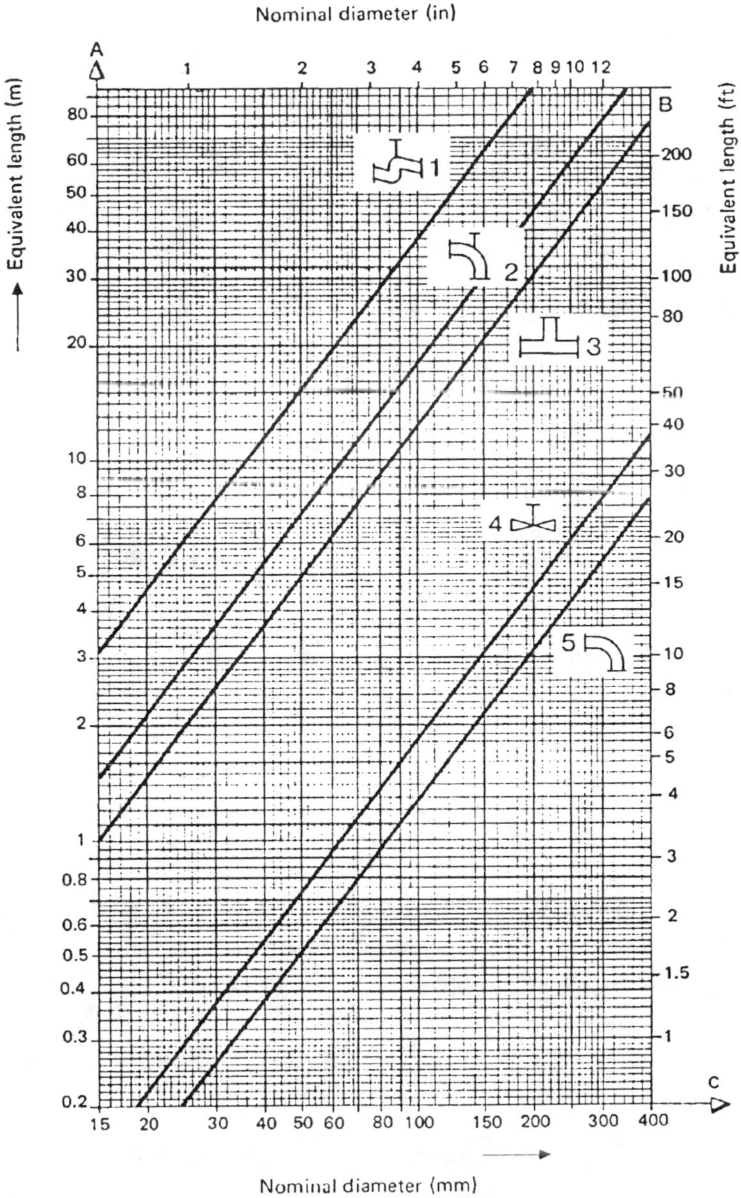

Figure 9.4 Nomogram. To convert a restrictive device in an air line to the equivalent length of pipe, find the nominal size of the globe or needle valve (1), angle globe or needle valve (2), tee (3), gate valve (4) or elbow (5). When the nominal size of the restrictive device is known, follow the vertical line to the heavy indicating line and then the horizontal line to the equivalent length in meters or feet.

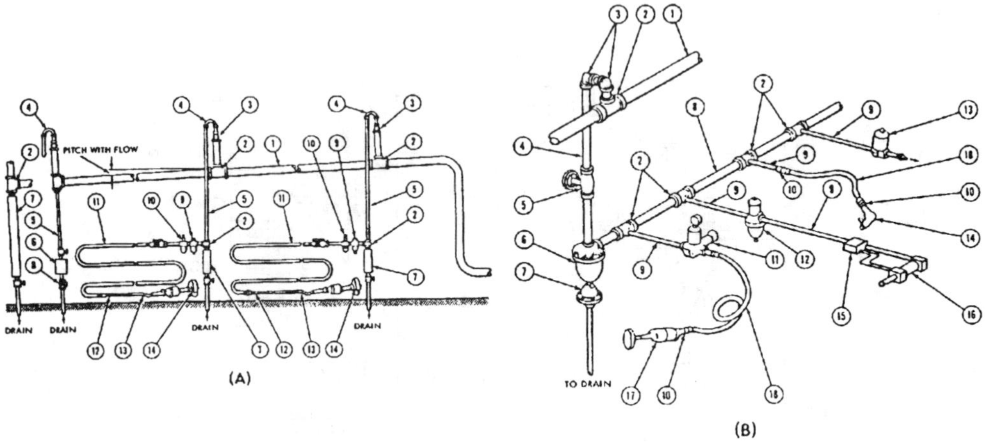

Figure 9.5 Suggested installation of piping and accessories for pneumatic portable power tool application: (A) individual down lines provided for each tool: (1) main air line piping; (2) tee; (3) branch main air piping; (4) wide pattern return bend; (5) drop line; (6) water trap or (7) water leg; (8) three-way valve; (9) filter; (10) lubricator; (11) feeder hose; (12) reducer; (13) flexible hose whip; (14) portable grinder and (B) common drop line supplying a number of tools: (1) main air line piping; (2) tee; (3) elbow; (4) drop line; (5) gate valve; (6) filter; (7) automatic water trap; (8) feeder line; (9) individual tool feeder piping, rigid; (10) quick-disconnect coupling; (11) regulator-lubricator unit; (12) filter-lubricator unit; (13) lubricator; (14) blow gun; (15) valve; (16) cylinder; (17) portable grinder; (18) air hose.

such an installation, it is necessary to prevent or at least minimize the carryover and discharge of compressor oil.

TRAPPING THE WATER AND COMPRESSOR LUBRICANTS

The feeder line of Figure 9.5*B* is slanted towards the filter and water trap. If long distances are involved, it is recommended all air drop lines bringing air to working devices be taken from the top of the feed pipe to avoid the possible flow of water to individual tool piping, hose or tubing. *Drain* drop lines should be connected to the *bottom* of the supply lines.

LOW LEVEL INSTALLATION FOR ACCESSIBILITY

Air-line filters, pressure regulators, lubricators and the associated piping, fittings, etc. should receive particular attention regarding installation details and location. Figure 9.5 illustrates a low installation. Filters will more likely be cleaned and oil cups in lubricators will more likely be filled if they are accessible to maintenance personnel.

HIGH LEVEL INSTALLATION HAS SOME ADVANTAGES

Figure 9.6 illustrates an elevated installation. The choice will depend upon plant structural details, piping elevation, and operating requirements. The low level installation (Figure 9.5) provides better accessibility and visibility. It has the disadvantage in some instances of occupying otherwise valuable factory floor space, and is more susceptible to damage or being tampered with by unauthorized or inexperienced personnel. The elevated installation of the filter-regulator-lubricator assembly (Figure 9.6) has the advantage of less interference with factory floor space and better maneuverability of air hose and portable power tools for certain types of applications. The installation shown in Figure 9.6B has the advantage of requiring only one filter for two tools for operations which do not require pressure regulation of the tools. The higher installation also limits, to a greater degree, the amount of air hose lying on the floor or dragging along the floor. It has the disadvantage of more difficult accessibility for installation and maintenance, in pressure regulator adjustment and in filling the lubricator.

PIPING AND AUXILIARIES FOR MACHINE TOOLS

Installation of piping and auxiliary air line equipment for machine tools is a relatively simple matter. Most machine tools are equipped with the auxiliary air equipment already suppled and installed by the machine tool manufacturer. As a result, all that is required is the connection of an air line to the machine, taking the precaution of providing moisture trap and/or ther provisions to prevent water or condensed compressor lubricant in the air supply line from migrating to the machine tool pneumatic system.

Should a machine be shipped without all of the pneumatic equipment in place, an air-line is usually installed overhead with a drop to the machine bearing suitable water exclusion features. The auxiliary equipment is then attached to the feeder line extending from the drop line (Figure 9.7). Factors for drainage and maintenance are the same as for portable pneumatic tools. Installation on machine tools should be made so that units are visible and easily maintained. Adequate air flow must be considered for machine tool actuators so that the machine tool response is not adversely affected. This may be provided by installing a suitably-sized air receiver close to the devices which have large air flow demands on an intermittent basis.

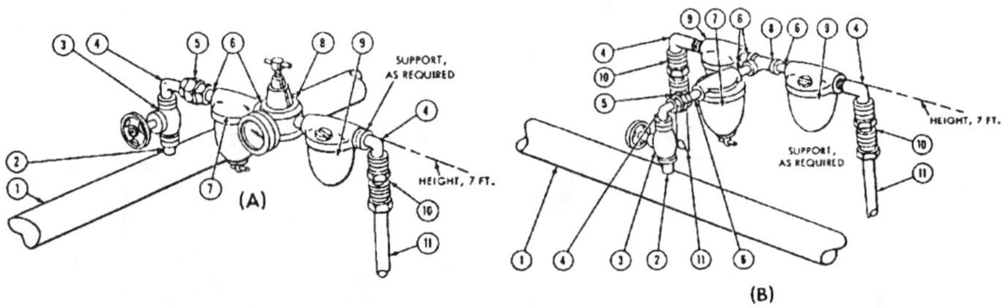

APPLICATION SIZES

FILTERS - REGULATORS - LUBRICATORS

SINGLE (BASIC) OUTLET:

C.F.M. REQUIREMENT OF TOOL*	N.P.T. PORT SIZES, IN.
1 - 6	1/4
5 - 15	3/8
12 - 50	1/2
45 - 90	3/4
80 - 140	1

*BASED ON TOOL MANUFACTURER'S SPECIFICATIONS

DUAL OUTLET:

	N.P.T. PORT SIZES, IN.†
FILTER	LUBRICATORS
(1) 1/2	(2) 1/4
(1) 1/2	(2) 3/8
(1) 1/2	(1) 1/4 & (1) 3/8
(1) 3/4	(1) 3/8 & (1) 1/2
(1) 3/4	(2) 1/2

† SIZES DETERMINED BY C.F.M. REQUIREMENT OF TOOL - SEE SINGLE OUTLET

(C)

Figure 9.6 Overhead installation of piping and accessories for pneumatic portable power tool application, designed for conveyorized assembly line operations: (A) single outlet: (1) branch main air line piping; (2) welded-in take-off, 3/4-inch pipe minimum; (3) gate valve; (4) elbow; (5) union; (6) nipple; (7) filter; (8) pressure regulator with gage, as required; (9) lubricator; (10) quick-disconnect coupling; (11) hose to tool. (B) dual outlet, where pressure regulator is not required: (1) branch main air line piping; (2) welded-in take-off, 3/4-inch pipe minimum; (3) gate valve; (4) elbow; (5) union; (6) nipple; (7) filter; (8) tee; (9) lubricator; (10) quick-disconnect coupling; (11) hose to tool; and (C) application sizes for filters, regulators and lubricators. (Courtesy Ford Motor Company)

LINES BETWEEN CONTROL AND ACTUATOR

Air cylinders cycling rapidly tend to heat up - depending on the length (and volume) of the air flow line between the air valve and the cylinder. If the air delivered to the cylinder is trapped, so that it's compressed and

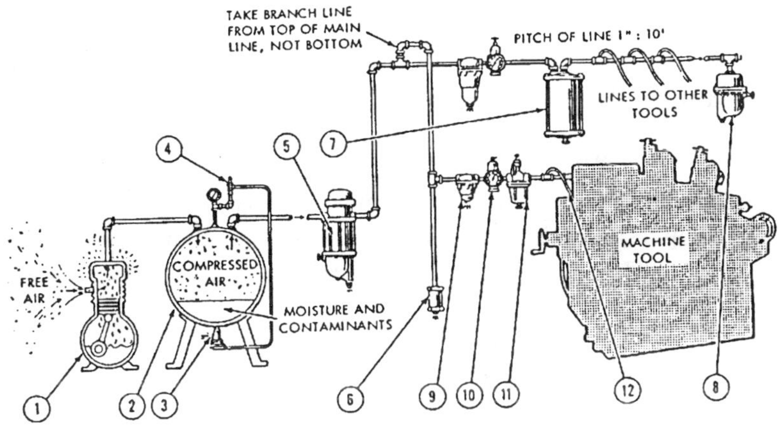

Figure 9.7 Pneumatic system designed to remove moisture, dirt, and other contaminants at various locations. (1) compressor, (2) receiver tank, (3) automatic tank drain, (4) governor, (5) separator-filter unit, (6) drop leg drain, (7) air filter, (8) sump drain, (9) air filter, (10) air pressure regulator, (11) lubricator, (12) air hose to eliminate pipe leaks caused by machine vibration. (Courtesy Wilkerson Corporation)

expanded but not exhausted, there can be significant heat generation which adversely affects lubrication, and oxidizes oils resulting in sludge deposits in the cylinder. This localized heat can also adversely affect seals. Keep air lines between control valves and the actuator short to avoid this potential loss of performance and system deterioration.

SUMMARY

Significant savings can be anticipated if provisions for emergency supply connections to a factory piping system are provided during the initial installation. High level piping saves floor space and minimizes damage to the pipes. Low level piping is easier to maintain and make connections for tools that are used intermittently.

Feeder lines should be pitched down toward water drains. Drain drop lines should be connected to the bottom of the supply lines. All air drop lines used to bring air to a working device must be taken from the top of the feed pipe to avoid the possible flow of water to the individual tool piping, hose or tubing.

REVIEW QUESTIONS

1. What type of device will protect a remote drop leg from accumulated water?

2. What is the purpose of the nomogram of Figure 9.3?
3. What causes compressed air to lose water as it moves through the piping system?
4. What are the two key factors to be considered when designing a drop leg?
5. What are the advantages to low level installations?
6. What are the advantages of high level installations?
7. Why is it desirable to keep air lines short between valves and actuators?
8. How can an adequate supply of air be provided for machines requiring intermittent high flows?
9. Which way should the lines slant in a typical factory installation?
10. What is pipe line resistance created by elbows, tees, valves, etc. normally related to when calculating circuit efficiency?

10

CONDITIONING THE AIR

PROPER CONDITIONING OF AIR IS A PRELUDE TO GOOD PERFORMANCE

Conditioning of the compressed air may include removal of moisture and solids, reducing the temperature and adding or removing oil or other lubricants. Proper conditioning of the air can be a significant deterrent to system and machine downtime. While compression of the air is a crucial first step, cleaning and drying it is a very important part of the good health and long life of the pneumatic power transmission system and a key to *Zero Downtime Pneumatics.*

Oil, dirt and water in the air wouldn't be too much of an operating problem if all the valves were poppets or other types that may seem to thrive on contamination, and if all the working components were built to work on wet, sludgy air. But those days (if they ever really existed) are disappearing. For example, modern close clearance spool valves for pneumatic logic quickly become victims of oil films. Oil in the exhaust brings on the government inspectors. Synthetic lubricants in plant compressors can carry over and swell certain seal materials and craze some types of plastics, especially filter bowls, causing early failure.

Air preparation for pneumatic systems must be far better than ever before. The market for filters, aerosol coalescers, dryers, refrigerant coolers, vortex coolers and filter, regulator, lubricator combinations is growing. Some recommend that exhaust mufflers and re-classifiers (to condense and coalesce oil vapors) not only capture the oil out of the exhaust, but return it to the reservoir.

Two processes are employed in the conditioning of the air. *Adsorbtion*, from the basic adsorb, means to collect (a gas, liquid or dissolved substance) in condensed form on a surface. *Absorption*, from the basic absorb, meaning to drink in, swallow up or engulf completely.

At one time it was enough just to aim the exhaust away from the operator, but not now. Today, federal regulations make it seem as if any air released into a breathing atmosphere must be purer than the air in a pine forest and no noisier than a swamp full of crickets! Tricks like exhausting air into passages within the body of the machine are clever, but it still must be "pure" - with no oil particles or other materials carried into the air which might be inhaled. Where no exhaust at all is allowed, as in some food processing or woodworking plants, the only answer is to pipe it away to a large enough coalescer.

AIR-LINE FILTERS

FILTER ELEMENTS

Filter elements of many types have been developed to remove contaminants from a compressed air system, thus reducing maintenance and the probability of damage to air-operated equipment. Some types of filter elements may or may not be suitable, depending on the filtration requirements.

CLOTH OR FELT-TYPE ELEMENTS

Filter elements employing cloth or felt are designed and manufactured to remove particles smaller than 5 microns (one micron equals 0.00003937 inch). Some elements are designed so they can be manually cleaned, though normally they are discarded. They may be more effective in removing small oil particles from the air stream than porous stone and sintered-metal types. This from the standpoint of new or freshly cleaned units which have less drop than other type elements with comparable ratings in particle-size removal. Wire mesh filter elements are used where a larger particle-removal rating is acceptable; for example in the 50 to 100 micron range. Impregnated paper and sintered-bronze elements are also effective in these size ranges. The paper and cloth filters are usually cartridges which are discarded and replaced as needed. The sintered-bronze, porous stone, and wire mesh-type are often treated as being readily and economically cleanable. While the cleanable types have a higher replacement cost, it may be difficult to determine whether it is cheaper to use an element which can be cleaned and reused or to install a throwaway-type at a lower replacement cost and an assured clean element. Selection will depend greatly on the condition of the compressed air system and requirements and the devices used in the air system.

HIGH FILTRATION WITH MINIMUM PRESSURE DROP

When high filtration (removal of small contaminants) with minimum pressure drop is required, the felt or cloth elements are reasonable choices. If highly "efficient" filtration is not required and cleaning or replacement of the filter element is required only between long periods of time, then the cleanable sintered metal or screen-type elements may be a better choice.

ADVANTAGES AND DISADVANTAGES

Each type of filter element has certain relative advantages and disadvantages. Paper and cloth elements both have a tendency over a period of time to become saturated with water or other airborne contaminants, and also to deteriorate. Therefore, since the compressed air system would tend to become heavily laden with moisture, replacement of these elements may be required more often. They also tend to clog, resulting in the increased pressure drop.

WIRE-MESH ELEMENTS

Wire-mesh type elements are not usually capable of filtering particles under 40 microns. They are also subject to damage through handling while being cleaned. Sometimes they are bent or distorted from being dropped or struck.

SINTERED-BRONZE OR POROUS STONE

The sintered-bronze or porous stone filter elements load or clog readily, with a resultant increase in pressure drop and a reduction in flow. Elements of this type in the 10 to 25 micron range are difficult to clean, and it is almost impossible to remove all of the foreign matter that has impregnated the element. Much solid material may come out of the air stream including core sand from pipe castings, pipe scale, etc.

LIQUID REMOVAL

Filters used on air lines, in addition to filtering out undesirable contaminants, also remove water and oil sludge which forms in the air line. The housing of the filter unit guides the compressed air into the filter section in a circular or swirling motion so that a centrifugal action is set up. This "throws" the water or other contaminant particles to the inside surface of the bowl allowing them to drop to the bottom in a "quiet zone" (Figure 10.1). The quiet zone is created by a baffle plate at or near the bottom of the filter element. The collected material - including water or oil sludge - can then be drained from the filter unit. The unit may also be designed to include louvers which force the air flow into a circular pattern. A series of baffles may also be included to help extract the water from the air. These are located near the top of the filter element which is close to the incoming compressed air.

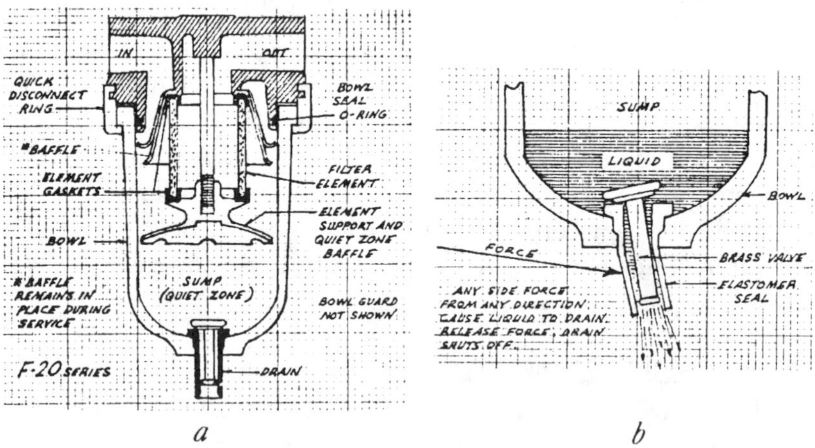

a *b*

Figure 10.1 Air line filter. (a) Sectional view. (b) Flexible draining device. (Courtesy Wilkerson Corporation)

These methods previously discussed will not remove all foreign material from the compressed air system. If this is desired, a more elaborate filtration unit is required which employs desiccants and/or oil-absorbing materials. Filter units can have automatic drains which will remove liquid as it collects in the bottom of the filter bowl. Some are provided with manual drains which can be opened as required.

TANK DRAINS, SEPARATORS AND DRYERS

In addition to air line filters, other important types of equipment are used to provide clean, dry compressed air. These include tank drains, moisture separators, air dryers and refrigeration systems. Each serves a specific function or need which is dependent upon circumstances, conditions and requirements. In some applications it is necessary that maximum removal of moisture, oil, dirt and micronic particles is accomplished. In other systems it may be necessary to remove only one or some of the contaminants.

MOISTURE SEPARATORS

One commonly-used method to trap water and other contaminants uses moisture separators with built-in traps to catch water and dirt (Figure 10.2). Where special drying of compressed air is necessary, in particular plant areas or departments such as food processing, instrumentation, or computer operation, one or more of the air drying methods may be advisable directly at the point where such air quality is needed. If the chief use of compressed air on a widespread plant basis demands pure air, it is usually more practical to condition the air at the compressor, the beginning of the system (Figure 10.3).

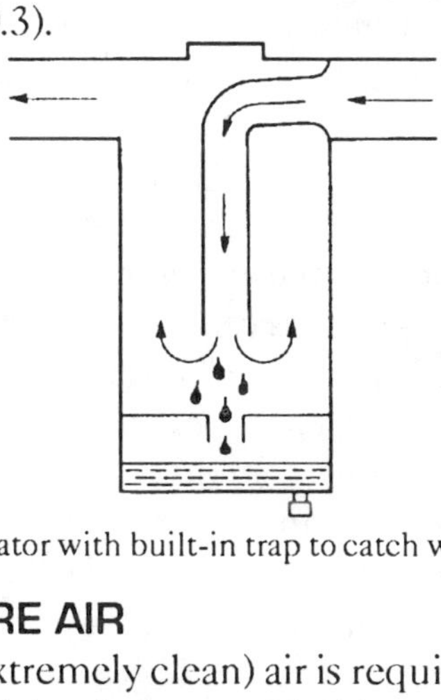

Figure 10.2 Separator with built-in trap to catch water and dirt.

INSTRUMENT PURE AIR

If instrument pure (extremely clean) air is required at certain points, its dew point can be sufficiently lowered by the use of small air dryers installed after an efficient oil-removing filter. If the compressed air leading

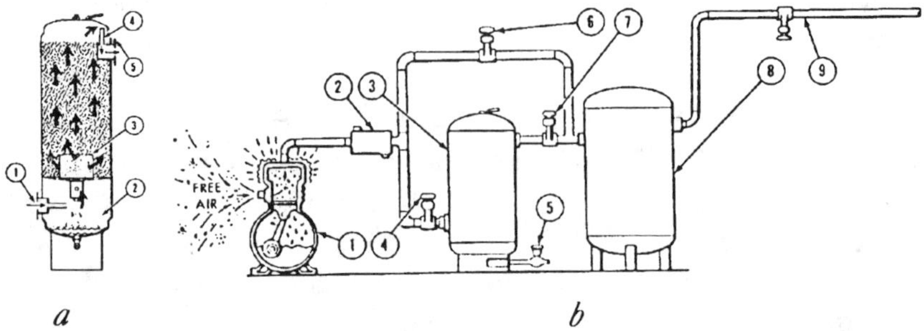

Figure 10.3 Where needs of dry air are anticipated throughout the factory. Use of compressed air dryer, employing deliquescent desiccants: (A) schematic view of dryer unit: (1) air inlet, (2) accumulator section, (3) stainless diffuser, (4) outlet baffle, (5) air outlet; (B) normal installation of dryer with system equipped with an aftercooler: (1) compressor, (2) aftercooler, (3) air dryer, (4) inlet shut-off valve, (5) drain valve, (6) by-pass shut-off valve, (7) outlet shut-off valve, (8) air receiver tank, (9) air supply line. (Courtesy Wilkerson Corporation)

to the device is occasionally shut off, such as overnight or during a weekend, then a dehydrator which contains a desiccant can be used. This requires manual changing when almost depleted. Manually-changed desiccant units are available in sets of two, equipped with an interconnecting hand valve. If only a minimum of downtime is allowable for changing the desiccant, the valve can be turned by hand to direct air into the stand by unit, permitting time for a re-charging and re-installation of the used desiccant unit. An automatic, self-regulating air dryer or a refrigeration-type unit that cools the air to its dew point can be used where manual changing of the desiccant is too time-consuming or costly.

DESICCANTS

Air dryers using the absorption process employ either one of two types of desiccants: (1) *Deliquescent* - the desiccant melts as it takes the water out of the air and then is drained off; (2) *Non-deliquescent* - the desiccant is regenerative and often can be used a number of times.

Three types of non-deliquescent dryers are available: (1) requires the manual removal of the desiccant from the dryer unit when the compressed air is shut off, and then it is replaced by a fresh supply. In some types, the used desiccant can be dried in an oven for reuse. (2) Uses a small portion of dry air to dry the standby non-deliquescent desiccant while the major portion of the dry air goes out the dryer's outlet. The air may be switched at short intervals from one tower of desiccant to another

(Figure 10.4). (3) Uses an electric heater to bake the non-deliquescent desiccant in one tower while the air is passing through the other tower. The air may be switched at long intervals from one desiccant tower to the other with an electric timer. In the type of dryer which uses a deliquescent desiccant, a quantity of desiccant is added when the air line can be shut off.

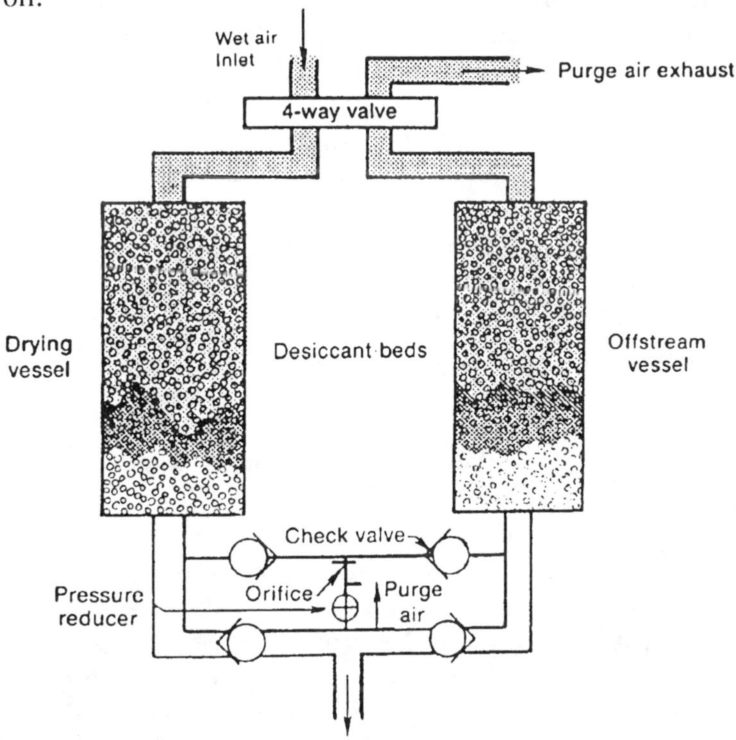

Figure 10.4 A small quantity of purge air can be used to dry the desiccant in the alternate vessel. (Courtesy Penton/IPC Publishing)

An afterfilter is desirable and may be required in conjunction with a desiccant dryer to remove desiccant dust from the downstream air. Where instrument-pure air is the major need and it is not desirable to have filters located at various points throughout the plant air system, or in such places as frozen food lockers and other places where low dew points are needed or perhaps required (low dew points often are desirable - but not necessary), large, deliquescent-type or refrigerator-type air dryers are used. To obtain still lower dew points for controlling atmosphere furnace gases, dehydration of natural gas, instrument air, gas transmission and storage environmental testing and maintaining dry communications systems, etc., the non-deliquescent desiccant units can be added where the gas is actually used.

TANK DRAINS

Contaminants from the outside which penetrate through the compressor intake air filtering system, plus moisture and compressor oils in various amounts, are subject to entering the receiver tank after passing through the compressor and/or after cooler. If the receiver tank has a manual drain and is not drained frequently enough it may accumulate foreign matter and water. As a result, some of the tank's air storage capacity can be lost, causing the compressor to run longer and more often. To combat this condition, an automatic tank drain may be used to purge the tank.

SEPARATORS

The use of receiver tank drains does not ensure the removal of all moisture and contaminants. Some contaminants such as fumes, micronic particles and compressor oil (especially that oil used to lubricate the piston and cylinder bore) may remain in suspension and be carried in the air stream to the main pipe lines. To protect machinery and equipment against the injurious effect of such contaminants, including the undesirable effects of some top cylinder lubricants - e.g. - those using certain phosphate ester additives, a single-unit, two-stage separator-filter equipped with an automatic drain may be used. An example of such a unit is shown in Figure 10.5. It is installed in the main line as far from the compressor as possible and ahead of the first branch line of the system. This removes moisture which is condensed by normal cooling in the transmission piping while the air is traveling from the compressor to the filter. It also removes moisture caused by air expansion and cooling when the air flows into the separator-filter unit itself.

In stage one of the separator-filter unit (Figure 10.5), the moisture and larger particulate contaminants are trapped and automatically removed as they accumulate. The relatively clean air then moves into stage two of the unit where the filter element removes micron-sized particles and most oil fumes. Since further cooling of the air can occur beyond the location of the main line separator, an air line filter is installed at each location where air is taken from the system.

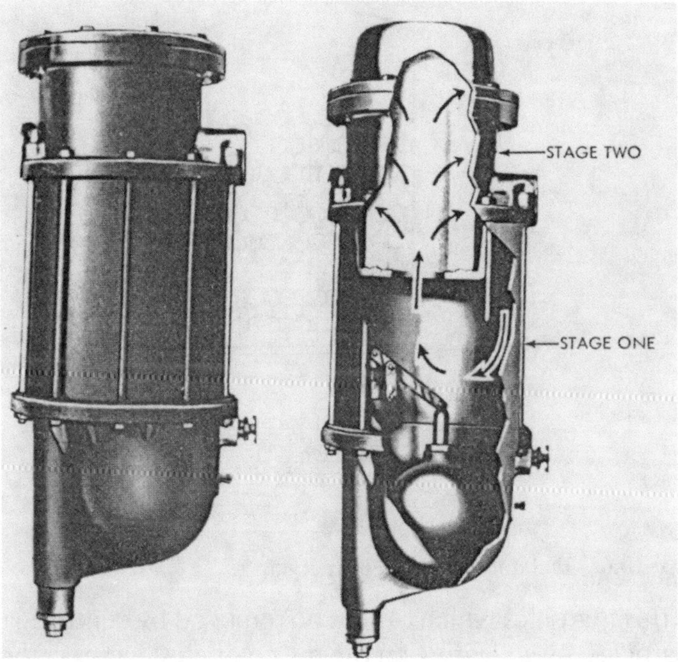

Figure 10.5 Separator-filter unit for main air supply line, equipped with an automatic drain. (Courtesy Wilkerson Corporation)

AIR DRYERS

An example of an air dryer which uses a deliquescent desiccant is shown schematically in Figure 10.6. The unit serves both as a chemical absorption air dryer and a separator. The unit functions as follows: air from the aftercooler enters at the inlet and condensed moisture and foreign matter collects in the accumulator area below the deflector plate. The moisture and foreign matter are drained from the lower segment. The compressed air, containing only suspended vapors and very small particles at this point, passes upward through the inner chamber into the desiccant bed. The chemical attracts and collects the moisture to the surface of the beads. When saturated the beads slowly dissolve, creating an emulsifying action of the water and other matter. As saturation is exceeded, the solution thus formed flows downward through the integral baffles where the upward-moving air stream causes the solution to disperse in a smoke-like mist which is extremely hygroscopic (absorbs moisture). After reaching this stabilization point, the created mist has also an affinity for attracting

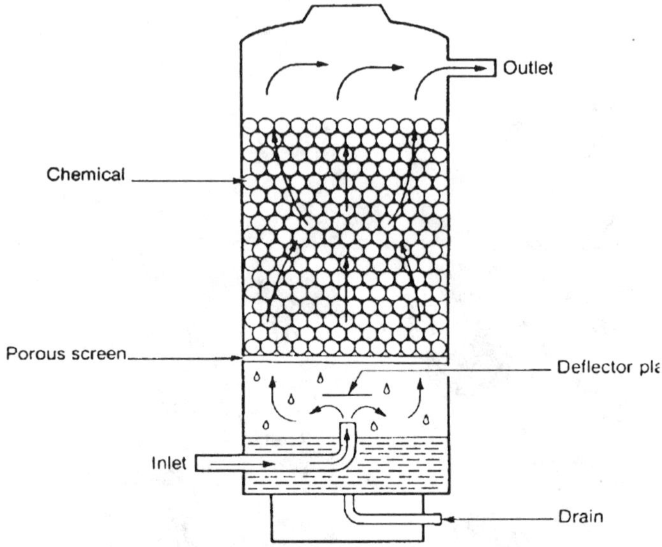

Figure 10.6 Deliquescent desiccant dryer.

airborne dust particles which cannot be removed by centrifugal action. These particles, by collecting on the mist particles, increase the weight and fall into the accumulator section for subsequent removal. As the air enters the outlet it has reached an instrument quality in terms of dew point. Additional data on the use of a refrigerant is covered in Chapter 11.

AIR-LINE LUBRICATORS

An air-line lubricator injects oil into the air stream to lubricate pneumatic cylinders, tools and control devices - such as directional valves - and system parts in contact with the flowing air. Examples are internal machine-tool parts such as bearings, cams, gears, etc., or to lubricate portable pneumatic power tools such as a grinder, nut-runner or drill.

Air-line lubricators store the lubricant and feed or inject it in controlled quantities into the air stream. The lubricant is usually stored in a transparent plastic bowl or a metal tank or larger volume device equipped with a sight-tube. The oil is fed into the air stream by means of a pressure differential, developed within the lubricator unit either by the venturi or the restriction principal. The pressure differential is used to elevate the oil from a bowl positioned below the piping center line or to force the oil past a check-ball when oil storage is above the pipe centerline.

INSTALLATION OF LUBRICATORS

The theory is that oil is added to the flowing air stream as minute droplets, small enough to be conveyed downstream to the tool. There can be a problem, however, because some of these droplets collect on the pipe walls, especially at bends and projections, and move down the line in liquid form. For this reason, lubricators working at pressures needed to do work should be installed above and as close to the tool as is practical, so that no pockets or low points will collect lubricant in the piping. Low velocity gives time for the air to increase oil droplet fallout and oversized piping *should not* be used.

The original *oil mist* devices developed by SKF for bearing lubrication and then sold to Alemite delivered oil carried in slowly flowing air very effectively at pressures *below* 5 psig. They require special devices to effectively coalesce the oil from the air stream.

POTENTIAL START-UP PROBLEMS

During initial installation of a tool likely to be damaged by a lack of lubrication, some oil should be added to the device and to the inlet piping to ensure that no damage caused by lack of lubrication will occur at the initial start-up. A lubricator adds oil in very minute quantities. If piping and equipment are initially dry, several minutes or even hours of operation may be necessary before an adequate coating of lubricant covers all surfaces.

LUBRICATION FOR AN AIR CYLINDER

Lubricator installations are usually straight-forward, but a problem can develop when a lubricator is used on an air cylinder application with a directional-control valve. Many lubricators act as a check valve, especially those creating a pressure differential by a resilient plug or spring-loaded valve. Reverse flow is restricted, so that installaton of the lubricator between the valve and cylinder is not desirable, unless slow piston return is also acceptable. Installation of a quick-exhaust valve may overcome this problem, but most of the lubricant will be lost at the exhaust port, without entering the air cylinder.

Installation upstream of the directional-control valve is most often the preferred method of installation and is generally satisfactory. If loss of oil at the exhaust port of the valve seems to be excessive, and if one direction of cylinder motion is slow while the reverse flow is rapid, the cylinder may not receive any lubricant on the slow traverse side. Another possibility for a single acting cylinder is that the lubricator may be installed between the control valve and cylinder, with a reverse-flow check valve parallel to the lubricator. Again, some loss of oil that has not passed through the cylinder may occur. Fortunately, single acting cylinders are rare. Perhaps the best solution, at least for a single-acting cylinder is to have two ports in the cylinder working end, using a standard five-port, four-way, two-position control valve to supply air to one port and exhaust air from the other. The lubricator should be installed between the valve and cylinder on the supply line, as shown in Figure 10.7. All lubricant exhausted from the valve in this system will have passed through the cylinder, thus reducing lubricant loss to a minimum.

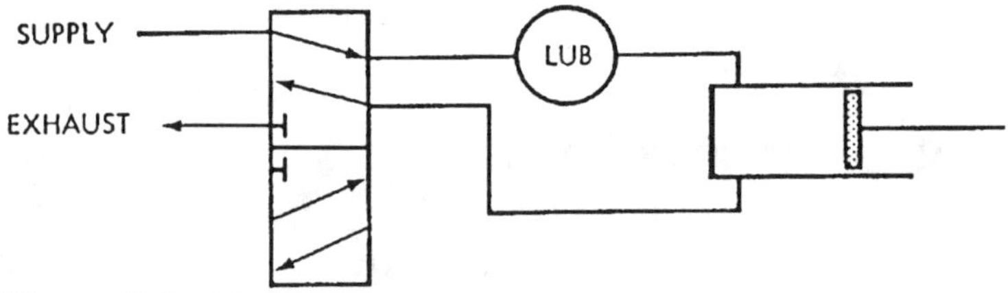

Figure 10.7 Five-way directional-control valve with lubricator as shown prevents excessive loss of lubricant for a single acting air cylinder installation.

Air cylinders cycling rapidly tend to heat up - depending on the length (and volume) of the air flow line between the air valve and the cylinder. If the air delivered to the cylinder is trapped, so that it is compressed and then expanded - but not exhausted - heat generation and lubrication problems can be expected. This heat will adversely affect the lubrication, oxidize oil and result in sludge in the cylinder which in turn can create seal problems, etc. At the same time, little or no oil may reach the cylinder. Air lines between the control and actuator should be as short as possible.

By far - most air cylinders used in fluid power circuits are double acting. If the traverse velocity in the extend mode is similar to that of the retract mode, lubrication is simplified as the air flow rate is then similar for both strokes (the effect of the piston rod volume is minor because the piston rod area is small compared to the bore area).

If the stroke velocities are significantly different, and you're using a single lubricator, install it so that it feeds the line leading to the rod end of the cylinder. That way, the bore of the cylinder is lubricated as well as the rod bearing. On a single-acting cylinder, try to lubricate the rod end (Figure 10.8). Set the feed oil for the fast end of a double-acting cylinder.

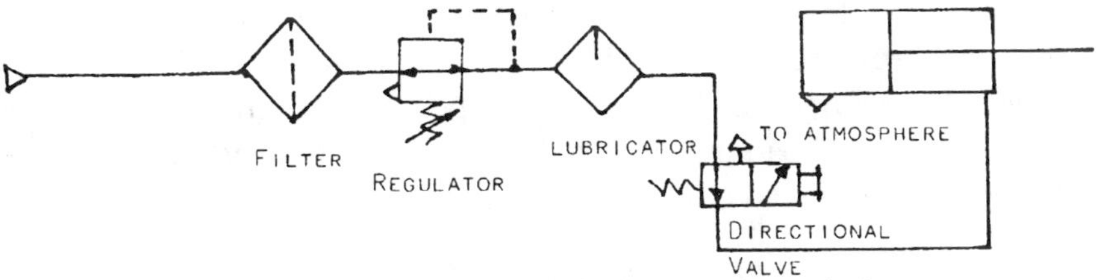

Figure 10.8 Lubricate the rod end of a single acting cylinder.

There *is* a problem with high oil flow rates into the cylinder, especially with cushioned cylinders. Oil coalesces and forms puddles in the cylinder - and at times the circuit acts like a hydraulic unit at the end of the stroke where the oil is trapped. The deceleration is very sudden compared to the original (non-oil flooded) setting. This can (and often does) result in a loud thump at the end of the stroke as the oil is compressed. The flooded cylinder moving at a fair speed cannot benefit from the cushion which had been set to function with compressed air. The liquid oil will result in a shock condition. Overfilling the lubricator (with oil) often results in problems shortly after the initial installation. If possible, use cylinders which do not need lubrication provided by an air-line lubricator. Many manufacturers make them available.

KEY LUBRICATION POINTS TO REMEMBER

1. Do not over-lubricate. This creates problems at cylinder stroke ends - and - when the oil gets spewed out at the exhaust - problems with OSHA will be significant.
2. Set lubrication flow for the fast end of a double-acting cylinder.
3. Set lubrication for the rod end of a cylinder where the traverse rates are slow.
4. Over-lubrication of cushioned cylinders will create vibration and shock problems. Adjust the lubricator to keep oil from flooding the cushioned ends.

CHARACTERISTICS OF LUBRICATORS

As compressed air flows through air line lubricators they develop a pressure differential. The higher pressure is applied to the surface of the oil in the reservoir and the oil is conducted by tube or wick to a point of lower pressure. To create the differential, some manufacturers use a simple venturi (Figure 10.9). Others use an adjustable restriction. Still others use a resilient restricting plug or spring-loaded valve. In addition to creating a pressure differential, most units provide a means to control oil flow, either by needle valve or wick exposure adjustment. Thus, the units which feature manually adjustable air restriction have two controls, while spring-loaded valve units have only one.

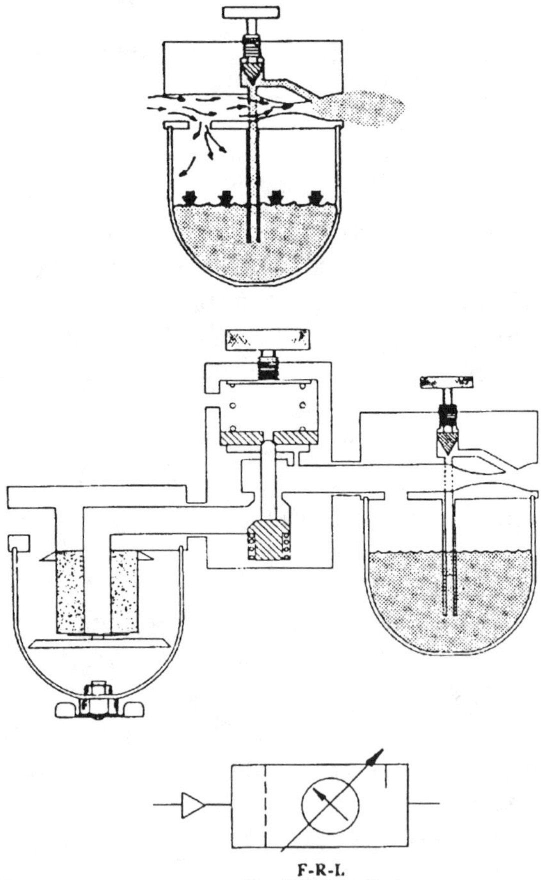

F-R-L
Simplified Symbol

Figure 10.9 Upper figure - air passing through simple venturi creates a pressure differential. Pressure is directed to surface of lubricant. Lubricant is forced through an adjusting mechanism to provide the desired quantity joining the downstream outlet air flow. Center figure - Typical filter, regulator and lubricator. Lower figure - F-R-L simplified symbol. (Courtesy Parker Hannifin Corporation)

HOW MUCH LUBRICANT?

The amount of oil being delivered is usually not too critical. Most tools need at least *some* oil. Flooding of the tool may however produce sluggish operation. To indicate delivery, a drop sight device is sometimes provided, and tool manufacturers specify the number of drops per minute that should be fed into the line leading to the tool. In some units not all the oil sighted in the lubricator's sight device remains in the air stream, while in others all of the visible oil goes into the line. Precise control of the oil flow should not be expected since all of the following items, in addition to the adjustment do affect the rate of oil seen in the sight device: air flow, air pressure, oil level, oil temperature and oil viscosity.

The simple, non-adjustable venturi types are restrictive in their range of operation. Low flow rates do not create enough pressure differential to produce oil flow, and large flow rates create excessive pressure drop and inefficiency of the system. However, for a narrow range of a particular flow, they may be the simplest, most reliable and inexpensive units.

Although lubricator units are available which will deliver oil with less than $^{1}/_{2}$-psi pressure drop, atomization is quite poor at the low air velocity thus created. An allowance of 2 to 5 psi pressure drop in the lubricator should be provided. One type of lubricator passes air through the bowl portion after oil is entrained, so as to drop out the larger particles and create a more uniform oil particle size to be carried out of the lubricator. However, for some applications, excessively fine droplets pass completely through the tool, creating an oil-laden atmosphere (mist) around the tool exhaust.

To determine the oil level in the lubricator, most manufacturers provide a transparent plastic bowl for the oil reservoir, except in larger sizes where typically a sight glass is mounted on the side of a metal reservoir. Some plastic bowls are unsuitable for use with fire-resistant lubricants of the phosphate ester-type, because of cracking and subsequent failure of the bowls.

LUBRICATOR OIL

Manufacturers of lubricators generally recommend a non-detergent petroleum-base oil, SAE 10 or lighter, for use with air line lubricators. Lighter oils are easier to atomize and since most air tools run cool, higher-viscosity oils are not necessary. As previously stated, some oils are not suitable for plastic bowls or other plastic or elastomeric components exposed to the air flow. Compatibility of a specific oil with seals in the lubricators and with the elastomers used in the seals of *all* devices which will be wetted by the lubricant should be established.

LUBRICATOR SYSTEMS

Metal reservoirs for oil are available with automatic-fill systems where large numbers of lubricators are in use. The large reservoir units are particularly desirable where daily oil consumption exceeds the lubricator reservoir capacity. But such a system may be difficult to install close to individual tools, and failure to deliver oil, or incipient flooding of the tool, may be more difficult to detect.

COMBINED FILTER-REGULATOR-LUBRICATOR UNITS

The use of a combined filter, regulator and lubricator is a frequent requirement. Combinations can be made locally of any two or three of these units, using pipe nipples with taper pipe threads. The combined units are also commercially available (Figure 10.10). Units which are combined in a single or coordinated housing are also available. Piping can be reduced and space may be minimized. The filter is first placed in the lineup since foreign matter may cause the regulator to malfunction. The lubricator is usually last, since oil mist or droplets tend to be deposited by turns or valves in the system. A gauge is usually included in the F-R-L as shown in Figure 10.9.

SUMMARY

Exhaust air from the pneumatic force and motion power transmitting system must be safe for the operating personnel. Contaminants must be trapped in filtering devices.

Careful analysis of costs may dictate use of cleanable filters for some installations and throw-away units in others. Allocation of time for service personnel may dictate the choice.

Air dryers may be necessary for many operations such as air gauging, instrumentation and certain packaging operations. The type of dryer will be dependent upon the service required and frequency of the cycling.

Lubrication must be tailored to the specific needs. Excessive lubrication can create contamination problems. Inadequate lubrication can lead to premature tool and machine wear. A clean, cool system - properly lubricated - is the prelude to *Zero Downtime Pneumatics*.

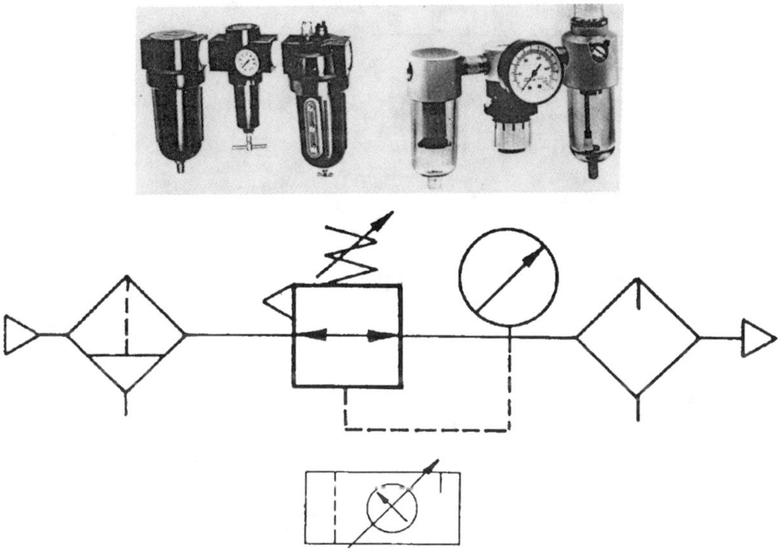

Figure 10.10 Upper figure - typical filter-regulator-lubricator unit. Center figure - ISO complete symbol. Lower figure - simplified symbol. (Courtesy Parker Hannifin Corporation)

REVIEW QUESTIONS

1. What are the four basic needs for conditioning compressed air for power transmission systems?
2. What is the major concern relative to oil mist in the exhaust systems?
3. Explain the adsorbtion processes.
4. How does adsorbtion differ from absorption?
5. What is the purpose of creating a swirling action in an air-line filter?
6. What is the purpose of a desiccant?
7. Why is it important to locate an air-line lubricator as close as possible to the device to be lubricated?
8. Why is it important not to use oversize piping with typical oil lubricators?
9. Should the lubrcator be installed ahead of the directional valve or between the directional valve and the actuator?
10. What type of problems are created by over-lubrication of cushioned cylinders?

11

HEAT TRANSFER

DISSIPATING HEAT OF COMPRESSION

The amount of heat generated in the compressing of air will dictate the type of heat transfer mechanism required to dissipate this product of the compressing function. The least complex device may be fins cast on the outer surface of the cylinder wall of a piston-type compressor or on the housing of a rotary compressor.

When staging is employed, the heat exchange may be accomplished with fins on the tubing between stages (Figure 11.1). The drive wheel may have blades cast into the structure to act as a fan to aid in the heat dissipation from the finned tube(s).

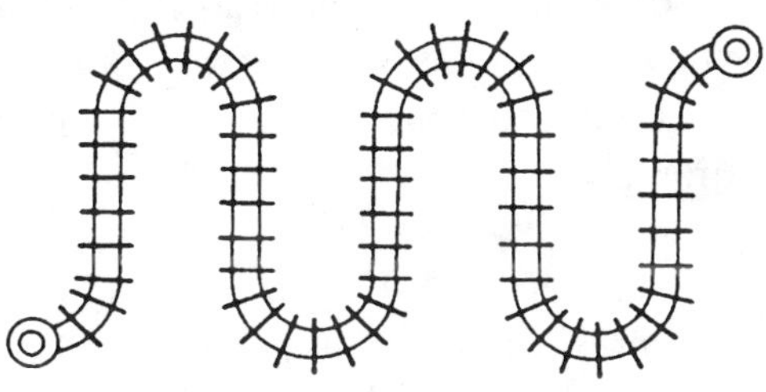

Fig. 11-1 The intercooler may be nothing more than finned tubing.

COMPRESSOR AIR

After being filtered to a large extent by the compressor intake filter, the air moves to the piston chambers. If it is a two-stage compressor, air is compressed with a consequent temperature rise in the first stage, passed through the intercooler where its temperature is decreased and then on to the second stage. In the second stage the air is compressed further and has another temperature increase.

THE INCREASE IN POTENTIAL ENERGY

The air at this point has increased in potential energy because it is compressed. But the air is hot and probably contains excess water vapor. In addition, air compressor outlet temperatures may run in the range of 400 to 500 degrees Fahrenheit (204 to 260 degrees C).

This air may contain some dirt not removed by the intake filter, and may carry some vapor picked up at the system air intake or while passing through the compressor. *Both are undesirable.* Heat is undesirable due to the expansion factor. If this hot air is passed directly into the pipes, they would lengthen from the heat. Contraction would occur upon compress or shutdown. This recurring process could cause joints to leak and reduce efficiency. Therefore, flexibility must be built in to compensate for thermal expansion. Instead of discharging the air directly into an air receiver for storage, the air is passed through an aftercooler.

THE FUNCTION OF THE AFTERCOOLER

As its name implies, an aftercooler cools compressed air after compression has been completed. This is accomplished by passing cooling water or air over the aftercooler chamber (Figure 11.2). Besides being the point where air cools, an aftercooler is also where additional dirt and oil vapor fall out of suspension, and a good portion of entrained water vapor coalesces out. The aftercooler must have a moisture separator, preferably with an automatic drain.

AN AFTERCOOLER CONDENSES WATER VAPOR

Specific volume per unit mass decreases as a gas cools. This change results in a decrease in the pressure of the gas. When air cools, its ability to carry water vapor also decreases.

A cool wind felt on a humid summer day is often the prelude to rain. When hot air mixes with cooler air, the temperature of the hot, moisture-

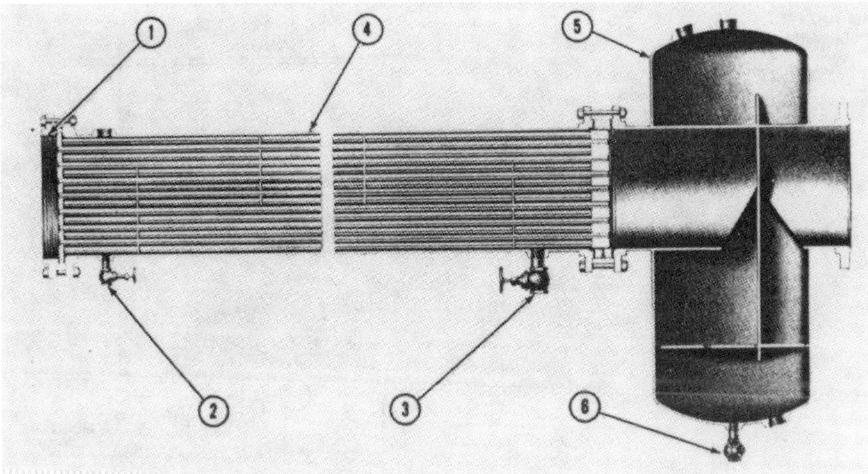

Fig. 11-2 Cutaway view of a one-stage aftercooler with a moisture separator. Compressed air is passed through tubes around which water flows in an opposite direction to the air flow. Compressed air velocity is reduced at the end of the aftercooler by a moisture separating chamber which reverses the direction of the air flow several times to trap out condensed moisture. (1) compressed air inlet, (2) cooling water outlet, (3) cooling water inlet, (4) aftercooler, (5) moisture separator and condensate trap, (6) condensate drain for connection to automatic trap. (Courtesy of Ingersoll-Rand Company)

laden air decreases and so does the air's ability to hold water. Condensation or "rain" results.

As a function of the aftercooler, compressed air and the water vapor it carries is cooled and its water content is made to condense and "rain". As air passes to the receiver, it has much less potential energy than when it entered. However, it is cooler, cleaner and holds less water. The air leaving the compressor is typically very humid. This high humidity requires an aftercooler to remove some of the water vapor.

The direction of water flow is opposite that of the air flow in a typical water-type aftercooler. A good aftercooler will cool the air flowing through it to within approximately fifteen degrees Fahrenheit of the cooling water temperature. It will also condense up to about 90% of the water vapor originally contained in the air as it enters the receiver tank (Figure 11.3).

Water and heat removal are both vital to an efficient pneumatic system. One thousand cubic feet (28.3 cubic meters) of air after compression can release as much as 1.4 quarts (1.3 liter) of water. Thus a modest size, 100 SCFM system could produce over 50 gallons of condensed water in

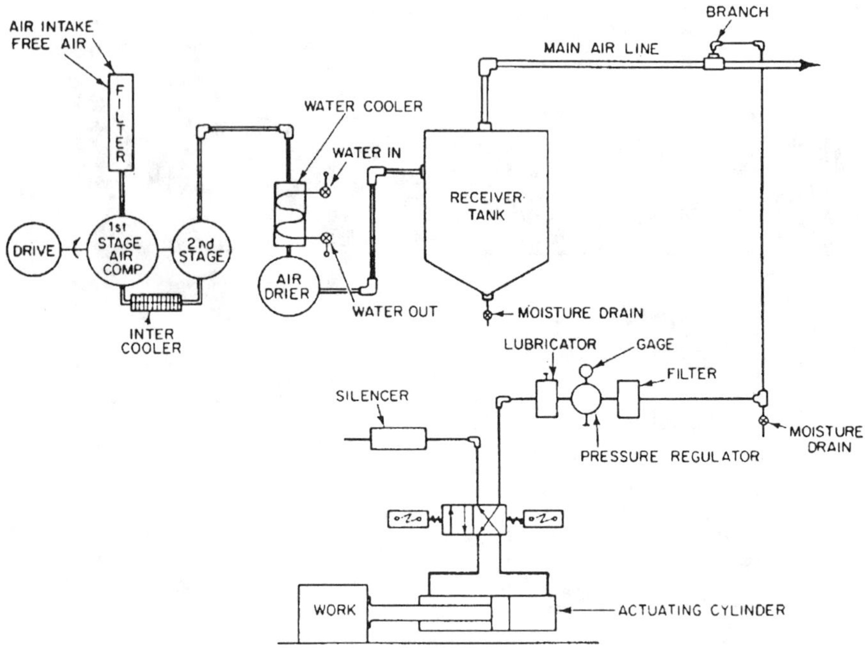

Fig. 11-3 Finned intercooler is supplemented by water cooler as the "aftercooler".

a single 24 hour day. If demands of a system require drier air, the compressed air can then be put through one of the following processes: a) Over compression, b) Refrigeration, c) Absorption, d) Adsorption or e) A combination of two or more of these methods. Let's study each one further.

OVER COMPRESSION

In the over compression process, the air is compressed so that the partial pressure of the water vapor exceeds the saturation pressure. The air is then allowed to expand, becoming cooler and drier. This is the simplest method, but power consumption is high. It is usually not used for large installations but is found most frequently in small systems. Therefore, it is not often found in major industry as are the other methods.

REFRIGERATION

When the temperature of air is lowered, its ability to hold gaseous water is reduced. This is what takes place in an aftercooler. The typical minimum air temperature attainable in an aftercooler is limited by the temperature of the cooling water or air. If extremely dry air is needed, a refrigerant-type cooler is employed.

When using a refrigerating-type unit, hot incoming air is allowed to exchange heat with the cold outgoing air in a heat exchanger. The circuit is shown in Figure 11.4. The lowest temperature to which the air is cooled is 32.4 degrees Fahrenheit to prevent frost from forming. This type of air drying equipment has relatively low initial and operating costs. Absorption and adsorption were discussed in Chapter Ten.

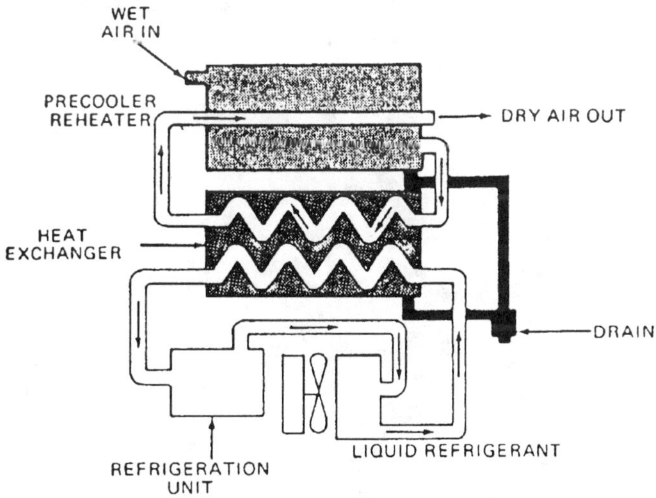

Fig. 11-4 A refrigeration unit employs a typical compressor with liquid refrigerant to supply the heat exchanger where the hot, moist air is cooled and dried. (Courtesy Penton/IPC Publishing Company)

SUMMARY

The efficiency of a pneumatic system is directly affected by the capabilities of the heat exchange devices. Few systems can function with hot, moist compressed air. Consequently, Zero Downtime Pneumatics are only attainable with a properly designed heat exchange system in perfect working order.

REVIEW QUESTIONS

1. What type of device is employed to permit dirt and oil vapor in the compressed air to fall out of suspension?
2. What is the direction of water flow in an aftercooler?

3. What is the lowest practical temperature that air can be cooled in a refrigerating type unit? Why?
4. What are the two key conditions resulting from the use of an aftercooler that are vital to an efficient pneumatic system?
5. Is a moisture separator essential to an aftercooler?
6. What is the least complex device to dissipate heat between compressor stages?
7. What is the purpose of the blades cast into the drive wheel of a compressor?
8. What mechanical damage can be created by change of heat levels within the piping systems?
9. What affects the ability of the compressed air to hold water?
10. If the demands of a system require drier air, what processes are available to provide the desired level? Name five.

12

EXHAUST CONSIDERATIONS

COMPLETING THE POWER CYCLE

Heat created in the compressing mode is dissipated so that moisture content is reduced. Rapid exhaust of air can reverse this process and a refrigerating effect can be encountered. An excess of moisture in the air can cause an icing effect as the air passes through a tool or pneumatic machine circuit. This can restrict the efficiency of the system whether it be a demolition tool, a packaging machine or some other pneumatic device operating at a rate requiring a high flow of air.

THE REFRIGERATING POTENTIAL

A manufacturer of lubrication devices had a grease pump with an exhaust configuration that produced a vortex spin (Figure 12.1). The pump would periodically freeze-up. An unusual phenomenon shut off the air on the exhaust side by freezing up the passages through the muffler. This product problem occurred about 1949 or 1950. The cure - realign the hole so there was no spin imposed on the flowing air.

At that time, some engineers thought the phenomenon could be used for refrigeration. A researcher by the name of Hilsch at Illinois Institute of Technology did work on it. He discovered the phenomenon by deliberately spinning the air using a disc with a hole in it at the air inlet to isolate the center flow from the outside flow. He did not try it without the diaphragm, but another researcher did. It also worked without the diaphragm but not as well. Maxwell knew of the phenomenon. His explanation - it must be a demon! So in the late 1940's it became know as Maxwell's Demon!

In some designs, such as that of Figure 12.1, the air flow passages may result in a vortex flow of the exhausting air. This will create a drop in air temperature of the air spinning out of the exhaust passage (B) - the drop in temperature can be sufficient to cause icing of water particles in the air stream. If there is a muffler attached, it may actually cause it to freeze up. The end which has its flow restricted or blocked (A) will rise in temperature, but this heat is carried away by conduction in the metal. The cure? Don't make parts where the air will be caused to spin!

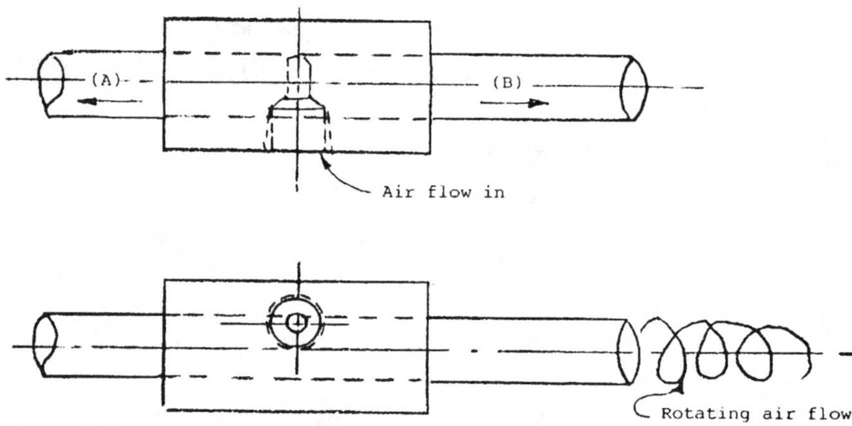

Figure 12.1 A column of spinning air can create a refrigerating effect.

A small test unit was fabricated such as that of Figure 12.1. With a flow of air at the inlet and a finger over (A), that end warmed up and (B) cooled. By holding a finger at (B) with (A) open, that end warmed up and (A) cooled.

LUBRICATION DISPERSION

A sluggish cylinder or pneumatic motor may result from excess lubrication. This may be compounded by oil or oil mist being exhausted to the surrounding atmosphere where it can damage human lungs or the factory product. One solution is to use an oil coalescer in the exhaust line.

Signs of excess oil in the exhaust system should trigger an immediate inspection of the lubrication system with adjustments to correct the fluid or oil mist loss. It is both an economic loss of lubricant and a potential health hazard.

THE MUFFLER

Excessive noise and/or other considerations may be the reason for installing a muffling device in the exhaust line of a pneumatic component. Mufflers are usually desirable where compressed air is exhausted into the atmosphere from air-operated equipment. Exhausting air may make a noise ranging from a "hiss" to a loud "bang". This may result in decreased worker efficiency through mental fatigue, temporary loss of hearing, irritation or other job related ailments. This, in turn, leads to decreased concentration on the operator's duties resulting in accidents and lower production.

Many muffler units are available that are effective in reducing air noise and blast. These units reduce the air velocity, change its direction, and dampen the vibration. They are available in a large range of sizes and types of construction. Commercial sizes range from 1/8-inch to 2-1/2-inch NPT. Some machines have the muffler assembly custom designed as an integral part of the machine structure. Small internal combustion engine mufflers are also available which can be used for pneumatic exhaust applications.

Figure 12.2A illustrates a 2-1/4-inch diameter x 4-inch long exhaust muffler designed for heavy-duty industrial applications in conjunction with such facilities as air exhaust valves, vacuum pumps, air compressors, air motors and air boosters. Air entering the muffler rebounds from a baffle or deflector plate and collides with incoming air, flowing to the exhaust port through a network of holes.

By exhausting the air into a muffler the air flow rate is decreased by restriction. The pressure caused by this velocity change is referred to as *back pressure.* As might be expected, the noise level of a muffler increases with the air flow rate which is a function of the flow resistance of the unit. This is reflected in back pressure shown in Figure 12.2B. When selecting a muffler, it is important that the flow required does not create a back pressure that adversely affects the operation of the system.

When a double-acting cylinder is actuated by a four-way, five-port, directional control valve (with dual exhaust ports), it is often desirable to

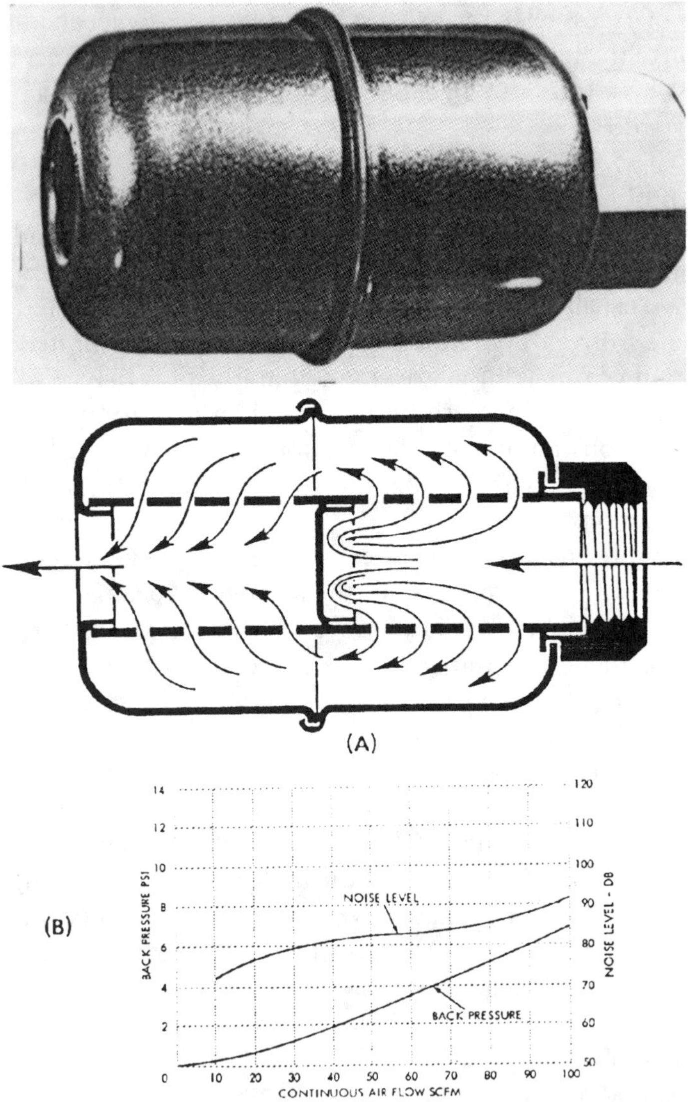

Figure 12.2 Compressed air exhaust muffler: (A) muffler for heavy-duty applications. (Courtesy Auto-Pontents, Inc.) (B) relationship of noise level and back-pressure to air flow for a 1/2-inch NPT compressed air exhaust. (Courtesy C. W. Morris Co.)

somewhat regulate the piston speed with flow-control devices installed in the valve exhaust ports. Combination "flow-control" mufflers can often be effectively used. A widely used type of "speed-control" muffler contains a porous, sintered-metal element to reduce air noise and blast (Figure 12.3). The terms "flow-control" or "speed-control" are used to describe a relative result of pneumatic restricting devices. Adjustable flow control mufflers are available as well as the restrictive devices mentioned. Motion control is achieved only when the inlet air pressure and downstream pressure remains constant.

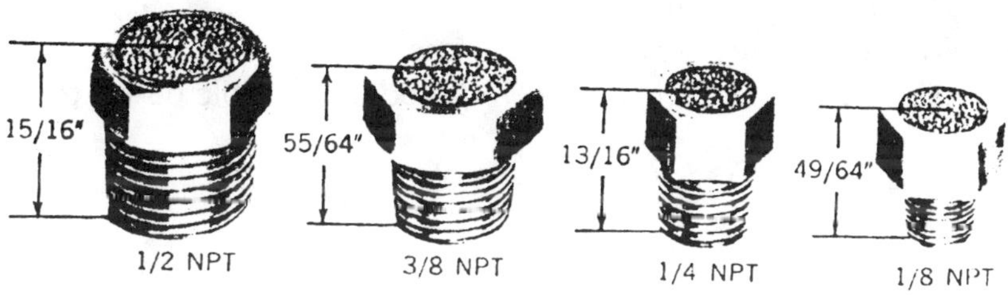

Figure 12.3 Mufflers or dust filters. (Courtesy Air-Mite, Inc.)

Manufacturers of portable air tools usually design the tools' air exhaust so that a muffling effect is obtained. It is not recommended that the user add any type of muffler. An exception to this may occur in the combined application of a number of air motors in a multiple spindle portable power tool.

SUMMARY

Expanding air can create a refrigerating action. Design care is needed to prevent icing of the exhaust air. Exhaust air must be monitored frequently to ensure clean dry discharge into the atmosphere. Moist and/or oily residue may indicate need for servicing the lubricating devices.

A muffler can restrict the discharge sound level from the pneumatic devices, minimizing operator fatigue. An excessive restriction can reduce the efficiency of the pneumatic power transmission system.

REVIEW QUESTIONS

1. What effect is created with the rapid exhaust of air?
2. Can excess lubrication create a sluggish cylinder or pneumatic motor? Why?
3. What type of adjustments are indicated with signs of excess oil in the exhaust system?
4. What type of job-related ailments result from inadequate sound level control for a pneumatic system?

5. What is the function of back pressure in a muffler designed for pneumatic service?
6. What two factors are required for motion control when using restrictive devices in a pneumatic system?
7. What phenomena creates icing of exhaust air? Will restrictions created by the pneumatic exhaust affect efficiency of the power transmission system?
9. What is the purpose of an oil coalescer?
10. Are external mufflers usually required for portable power tools?

13

AIR LINES AND CONNECTORS

Major air lines from the compressor and receiving tank may consist of welded or threaded rigid metal pipe. The air drop, which is to be connected from the top of the major distribution line to avoid moisture collecting in the drop line, may be fabricated from metal pipe or tubing. As an alternate there is a major trend towards the use of hard or soft plastic conductors to reduce line friction and rusting. Plastic lines usually have a long life span and are easy to fabricate. However, many chemicals and/or sunlight can reduce the life of plastic parts.

FLEXIBLE PLASTIC TUBING AND QUICK CONNECT/DISCONNECT FITTINGS

The following material was provided by Nycoil Company, 57 South Avenue, Fanwood, New Jersey (a member of NFPA) and first presented in "Technical Update On Fluid Power" provided by the National Fluid Power Association and published in *Product Design & Development Magazine* published by Chilton Co., Radnor, Pennsylvania.

In the early days of pneumatics prior to World War II, compressed air connections were usually made with rigid pipe and rubber hose. The fittings were expensive and bulky. Some fine threads were used for connections, but most had NPT pipe threads or barbs and clamps.

As the technology in the air industry developed, the variety of fittings grew, particularly when bendable copper tubing became popular as a substitute for rigid pipe. Later in the post-war period, flexible plastic tubing was developed for use with pressurized systems. This development led to more types of fittings designed for use specifically with plastics.

Nowadays, the pneumatics industry has many connector options, including a wide variety of pipe, hose, and tubing in metals and plastics. There is an even greater variety of fittings and connectors to fit all kinds of applications.

Surprisingly, many of the materials and methods used a hundred years ago are still considered the best for their respective applications. For example, flexible rubber hose is still used on railroad car connections, and black iron pipe is still considered the best choice for permanent industrial plant air supply lines.

On the other hand, hardly a year goes by without manufacturers coming up with a number of innovations in air connection methods. Predictably, the trend has been towards the use of materials, designs, and methods that improve performance and at the same time, reduce overall costs.

Because the biggest cost ingredient is often labor, industry has been moving away from air connections that require labor intensive methods such as cutting and threading pipe, the flaring of tubing, and the crimping and clamping of fittings.

Many of these methods are still used in hydraulics, where they are necessary to contain pressurized liquids, but most air systems up to 200 psig no longer need such methods. In fact, some of the newest air connections require very simple skills for assembly and almost no tools.

FLEXIBLE PLASTIC TUBING

In today's highly automated and sophisticated market, there has been a rapid adoption of flexible plastic tubing for use with compressed air systems. The three most popular tubing choices are manufactured of Polyethylene, Nylon, and Polyurethane thermo plastic materials. Polyethylene is the least expensive and, consequently, the most commonly used for less demanding applications.

Nylon is also widely used because of its superior qualities. Nylon withstands higher working pressures up to around 250 psig, higher ambient temperatures, and is resistant to many chemicals; but it costs more than double the price of Polyethylene.

Polyurethane has lower working pressures and temperatures than Nylon but much greater flexibility and abrasion resistance. It costs a little more than Nylon and is more difficult to connect because of its soft, flexible nature. However, where temperature and pressure permits, Polyurethane is often chosen because it lends itself to compact installations where tighter radial bends are needed.

This use of flexible plastic tubing has prompted an evolution in the design of fittings. The first type of fitting used with plastic tubing was the compression fitting, made up of a brass body, brass ferrule, and brass nut, plus a brass insert for inside the tube.

When the nut is tightened, it compresses the ferrule to make a seal and grip. This wedging action causes the ferrule to seal against the body and bite into the tubing, forming a permanent seal. The deformation caused by the compression ferrule and insert placed within the tube tends to restrict air flow.

Different compression fittings have been developed, mainly with variations in the ferrule material, such as brass and plastic and with different shapes and lengths of the ferrule. Other less commonly used compression fittings that do not deform the tube are made with resilient seals instead of ferrules.

Compression fittings have the largest share of the compressed air connector market, and are therefore usually the least expensive. However, in the past few years as labor costs have risen along with the constant need to check systems, there has been a strong trend toward the use of push-in type fittings for plastic tubing. These fittings are often called *instant fittings* or *automatic locking fittings*.

PUSH-IN TYPE FITTINGS

When push-in fittings were introduced to the market more than ten years ago, they were very expensive, often more than double the cost of compressions fittings. As their volume has increased, however, the cost differential has narrowed significantly.

Although push-in fittings still cost somewhat more than compression fittings, they are viewed by many users as less expensive overall after considering labor costs for installation and repairs, plus costs for parts and inventory. In addition, instant push-in fittings provide the advantages of full flow, convenience of use, and re-usability. A compression fitting is only partially re-usable because ferrules are deformed and need to be replaced when removed.

Push-in fittings are re-usable because they can be easily disconnected and then reconnected by hand. This is important when testing a system for maintenance or installation purposes. Push-in fittings also provide air flow, offer space savings for compact design, and are self-contained with no loose parts.

Instant push-in fittings were pioneered in Europe. There are several manufacturers, both domestic and foreign, for this type of fitting serving the air power market. Several designs have evolved; all generally falling into five categories, as follows (Figure 13.1)

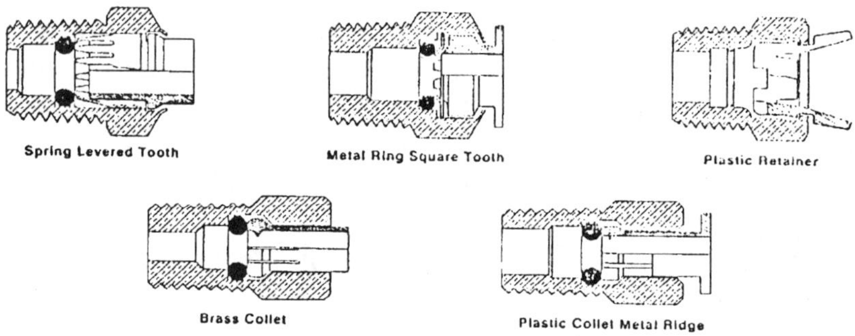

Figure 13.1 Instant push-in fitting designs. (Courtesy National Fluid Power Association)

1. SPRING LEVERED TOOTH

This design is equipped with 15 or more spring brass levered teeth that grip tubing. It includes an O-ring seal and release collar for manual disconnections. This type of fitting works with both soft and hard air tubing, including Polyethylene, Nylon and Polyurethane.

2. BRASS COLLET

This design has gripping ridges and is generally suitable for Nylon air tubing, which needs a hard thermoplastic gripping surface.

There are several styles available. The collets vary in shape with a different number of slotted ridges. They have anywhere from three to eight gripping ridges and have an O-ring seal.

The beveled shape of the body activates the collet, which grips against the tubing when it's pressurized. The fitting can be disconnected by pushing on the collet.

3. **METAL RING SQUARE TOOTH**
   This design, not suitable for soft tubing, includes a spring, six teeth, and an O-ring seal. It has a large plastic collar that releases the fitting.
4. **PLASTIC COLLET METAL RIDGE**
   This type of fitting, which is suitable for Polyethylene and Nylon, has a sharp metal ridge imbedded into a plastic collet usually having four to six sections. The plastic collet is used to disconnect the O-ring pressure seal. There is a large collar for finger disconnect.
5. **PLASTIC RETAINER**
   This design has pinch retainer ears to release and a beaded brass ring to seal. It is used for Polyethylene only. These five basic push-in fittings have varying features that warrant careful comparison before selection. Questions to ask include:
   - What is their ability to grip on hard and soft thermoplastic tubing?
   - What are their pressure and temperature ranges?
   - Are they suitable for use with water and other liquids?
   - What is the size of an individual fitting?
   - What was the material used to manufacture it: Brass? Plastic? Stainless Steel?

INSTALLATION AIDS FOR PLASTIC TUBE AND ASSOCIATED FITTINGS

Male pipe threads on some brass pipe connectors are pre-taped with Teflon to assure a uniform seal against leakage. Automatic pre-taping provides the exact amount of Teflon required. Precise application leaves the first thread exposed to eliminate the possibility of damage or downtime caused by excess tape tearing off, lodging in the circuit and blocking air flow (Figure 13.2).

To aid in releasing tubing in limited access areas, a special tool is available with pronged end to push in the release collar prior to pulling the tube out (Figure 13.3). A razor sharp blade as a cutter for plastic tube assures square, clean cuts on most plastic tubing (Figure 13.4).

Figure 13.2 Male pipe threads pre-taped with Teflon. (Courtesy Nycoil Company)

Figure 13.3 Tube release aid. (Courtesy Nycoil Company)

Figure 13.4 Tube cutter for plastic tube with razor-sharp blade to assure square, clean cuts on all plastic tube. (Courtesy Nycoil Company)

IF POSSIBLE AVOID PIPE FITTINGS

Elbows, reducers, nipples, tees, unions, wyes (y) and other fittings should be used *only when necessary.* Such fittings cause airflow restriction and increase friction (Figure 13.5). The result is increased pressure drop and loss of power at the air-operated device. Fittings with taper pipe threads are difficult to "aim" in the desired direction or to replace in tight quarters. Replacement with metal or plastic tubing and fittings is recommended. Standard iron pipe is normally used for major large diameter air lines. Extra strong or double extra strong is normally used for hydraulic service but may be used for mechanical strength in pneumatic systems where external forces may possibly damage the air lines. Note the typical application in Figure 13.6 showing the use of tubing for a simple control

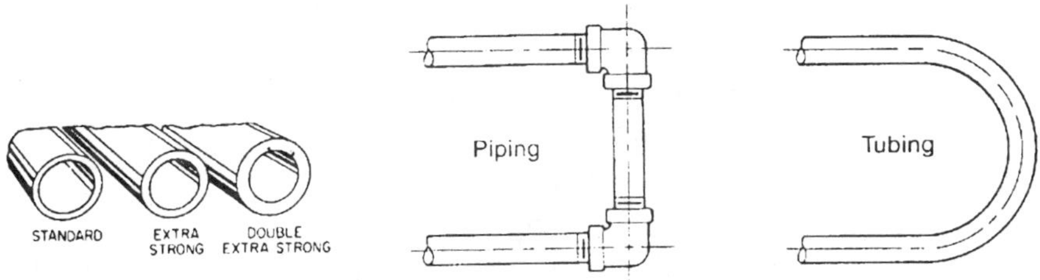

Figure 13.5 Tubing or hose eliminate extra joints and reduce frictional losses. Tubing is measured by its actual outside diameter (inch or metric system). Tube wall thickness will determine the resistance to flow as the inside diameter changes. Pipe is measured by a nominal bore size. The outer diameter remains constant for all wall thicknesses. Inside diameter varies with standard, extra strong and double extra strong. Hose is referred to by its inside diameter.

Figure 13.6 Typical application showing use of tubing. Requires but two fittings and joints. (Courtesy Southwest Industries)

line. Figure 13.7 illustrates the number of joints and fitting needed if typical threaded pipe fittings were used.

Loss of pressure due to air flow through fittings, elbows, bends, etc. may be expressed in terms of equivalent feet of straight pipe that would produce the same amount of pressure drop. For example, a 90-degree

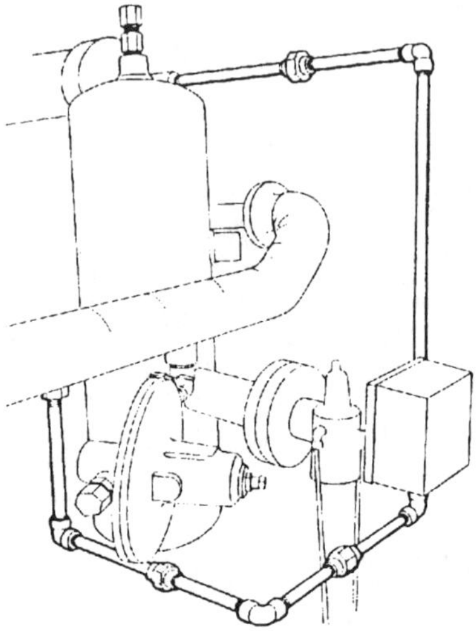

Figure 13.7 Schematic of similar type of application as in Figure 13.6 showing the six threaded joint and seven separate fittings required if piping were used.

short radius elbow for a 1-inch nominal pipe diameter has a rated friction loss equivalent to 1.6 feet of straight pipe. For a similar elbow for a 6-inch pipe, the friction loss is equivalent to 9.2 feet of straight pipe.

It should be necessary to use elbows only when branching off the main pipe lines from the upper surface or in feeder piping for tools or machinery using air. Nipples may be required when connecting to filters, regulators, lubricators or other auxiliary equipment. Reducers can be used when the feeder lines are large and the volume of air required for equipment operation is comparatively small. Sometimes larger-size pipe is used to reduce the velocity of the air, then used as close as possible to the air usage point, or at least at the beginning of the hose lines which feed the air-operated equipment. A union is used to permit two pipe lines to be connected or disconnected without rotating the pipe. Plastic, metal tube fittings or quick acting couplings can serve as a convenient disconnect means in place of the dedicated union whose only purpose is to provide the disconnect and connect function. When tubing is employed as the appropriate conductor it should be carefully laid out. Note the assembly of Figure 13.8. It is not always possible to make piping easily serviceable without careful planning. Note how difficult it would be to reach those fittings behind the front fittings. The installation in Figure 13.9 would be difficult to build without the advantage of tube fabrication.

Figure 13.8 It is not always possible to make piping easily serviceable. Care in initial design can help. Note how difficult (if not impossible) it would be to reach those fittings behind the front fittings. Tubing can be disassembled and reassembled with greater freedom than rigid NPT taper pipe thread piping and fittings.

Figure 13.9 It would be quite difficult to build the systems shown with pipe and fittings. The ease of bending and assembling metal tubing simplifies the installation of interconnecting lines. (Courtesy Automation U.S.A.)

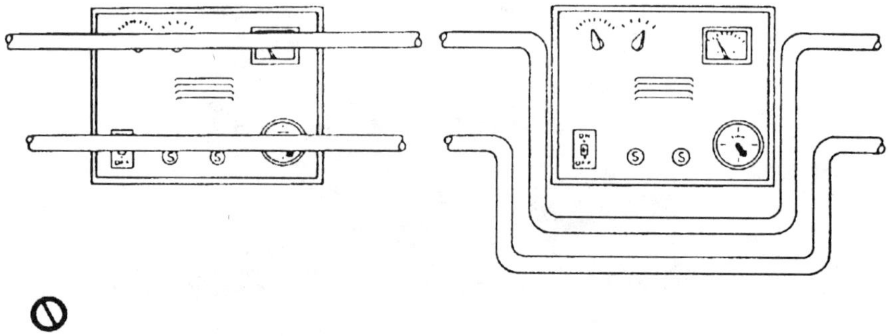

Wrong. Operating panel blocked. *Right.* Operating panel clear.

Figure 13.10 Watch for areas where adjustments must be made or components must be serviced.

It would seem unnecessary to even suggest that a good mechanic would carelessly run tubing in front of control panels. Yet, such things have been known to happen (Figure 13.10).

TUBE SUPPORT

Adequate support must be provided for tube and pipe lines. Mechanical damage can often be avoided if lines are properly supported. There is a temptation to use a low run as a bar rail for resting feet (Figure 13.11). Likewise, there is a greater temptation to use overhead lines located at a height of six or seven feet as a bus rail (Figure 13.12).

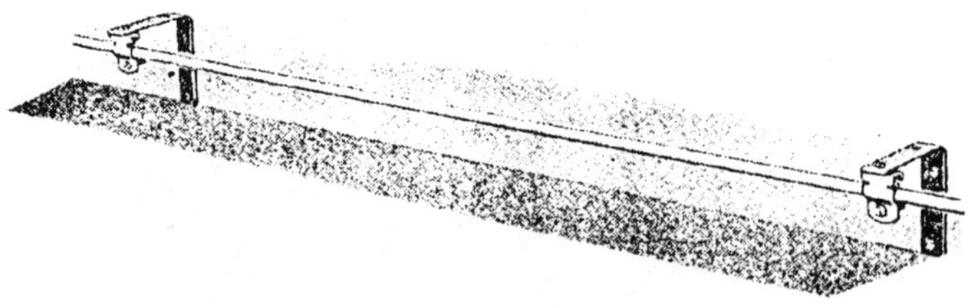

Figure 13.11 Bar rail.

Tubing should be supported in long runs to prevent sagging or low frequency vibrations (Figure 13.13). Tube size determines the spacing of supports for air lines. Generally, such supports should allow free axial movement of the tubing and support only the weight of the tubing (see Figure 13.14). All components should be mounted. The tubing should not be installed so that it must support the weight of valves, filters, regu-

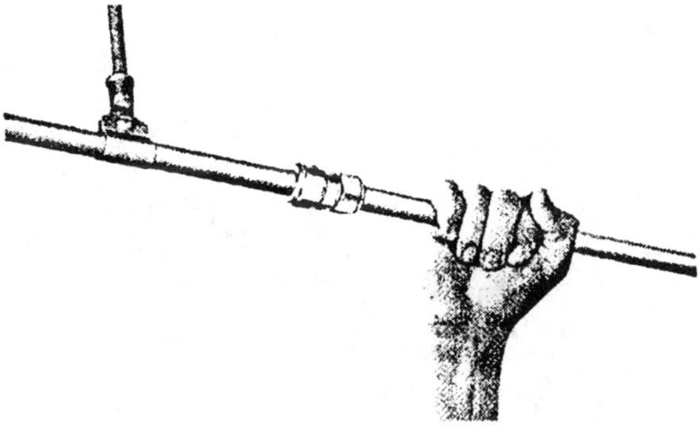

Figure 13.12 Bus rail.

Figure 13.13 Correct mounting.

Figure 13.14 Tubing supported in long run.

lators, etc., (See Figure 13.15). Tubing from 1/4-inch diameter through 1/2inch diameter should be supported at least every three feet. Sizes 5/8 through 7/8-inch diameter should be supported at least every five feet, and one inch and larger should be supported at least every seven feet. Tubing should be ganged vertically (Figure 13.16) rather than horizontally (Figure 13.17) to avoid collection of dirt, corrosives and contaminants.

Tube bending tools and instructions for their use are available from major suppliers of tubing, fittings and pneumatic systems. Most large suppliers offer "hands-on" tube design and fabrication.

Threaded metal pipe and/or metal tubing must be cut square with all burrs removed. Tube cutters do not remove material, but rather

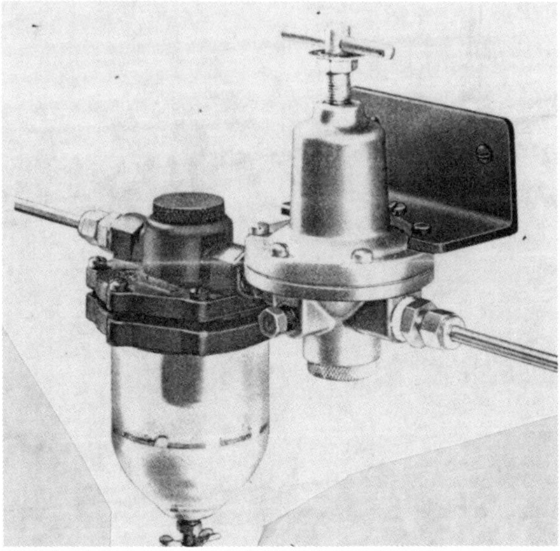

Figure 13.15 Lubricator with angle bracket support.

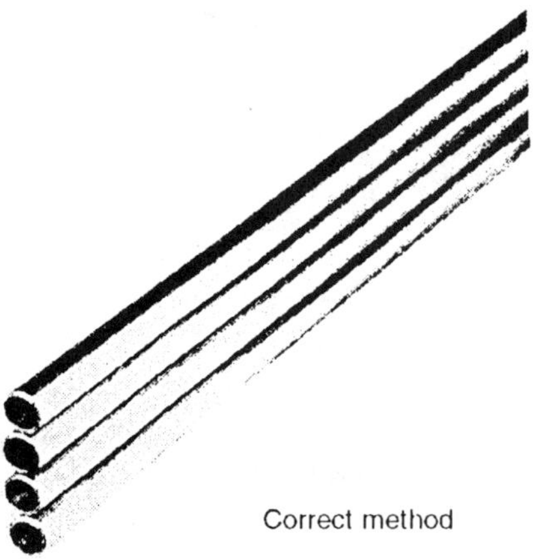

Figure 13.16 Vertical mounting of tubes avoids collection of dirt, dust and other foreign matter.

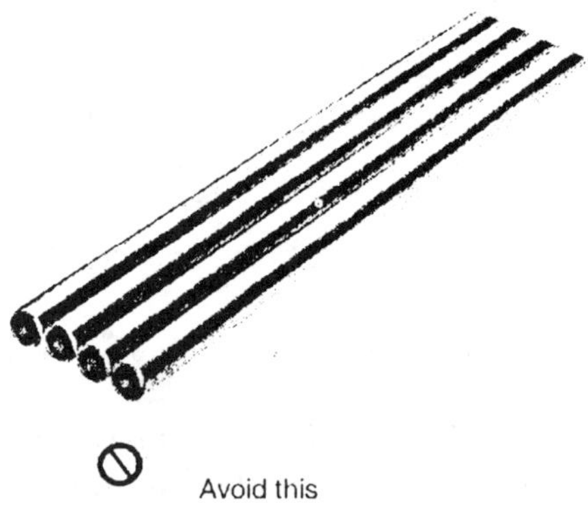

Figure 13.17 Avoid this.

"push" material aside and down. The duller the cutter wheel, the more material is displaced at the tube end. Depending upon the tube material, the cutting tool may increase or decrease the outer diameter at the cut end. In its worst case, raised material can make it difficult to insert the tube into precision tube fitting components. A dull cutter wheel also work hardens the area of the tube near the cut due to excessive pressure from back-up rollers. Figure 13.18 illustrates a typical tube cutter. Tightening the handle forces the cutting wheel deeper into tube wall (Figure 13.19). The cutter can either be rocked back and forth or turned completely around the tube (Figure 13.20). The tubing cutter assembly may have a marking on the handle to use as a guide when advancing the handle 1/8 turn (Figure 13.21).

Figure 13.18 Cutter for metal tubing. (Courtesy Crawford Fitting Company)

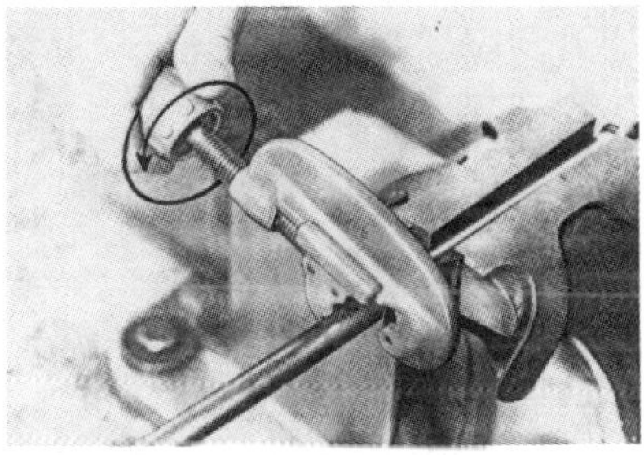

Figure 13.19 Tightening the handle forces cutting wheel deeper into tube wall. (Courtesy Crawford Fitting Company)

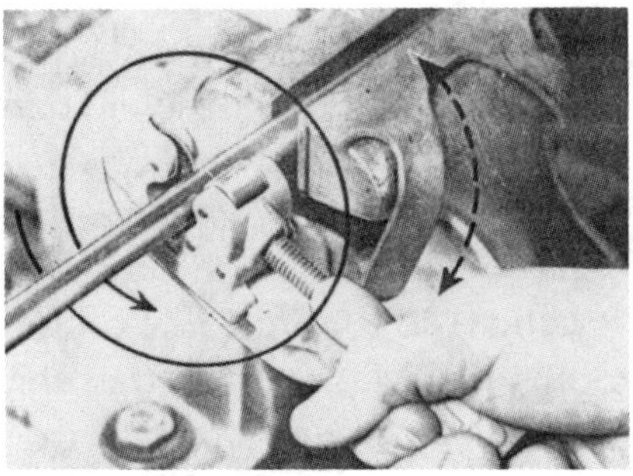

Figure 13.20 Cutter can be rocked back and forth or turned completely around the tube. (Courtesy Crawford Fitting Company)

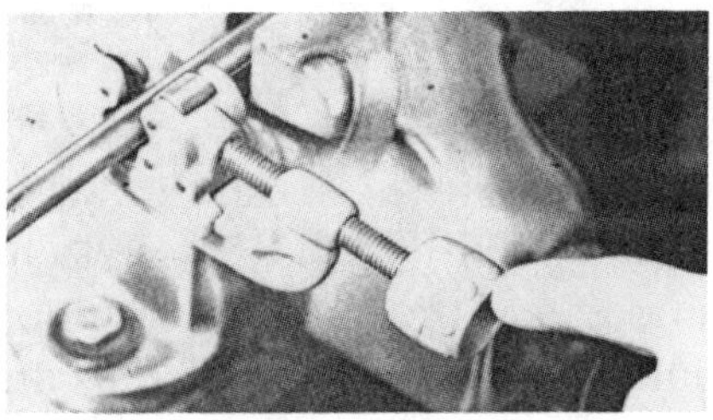

Figure 13.21 The raised portions on the handle of this cutter are placed in 1/8 turn increments. Use these as your guide when advancing handle 1/8 turn. (Courtesy Crawford Fitting Company)

TUBE CUTTING WITH A HACKSAW

If a tube cutter of the proper size is not available, a hacksaw can be used (Figure 13.22). Tubing should always be cut to length with a square cut. When using a hacksaw to cut tubing, guide-blocks are desirable to ensure a square cut and to keep the tubing from flattening out. Hacksaw

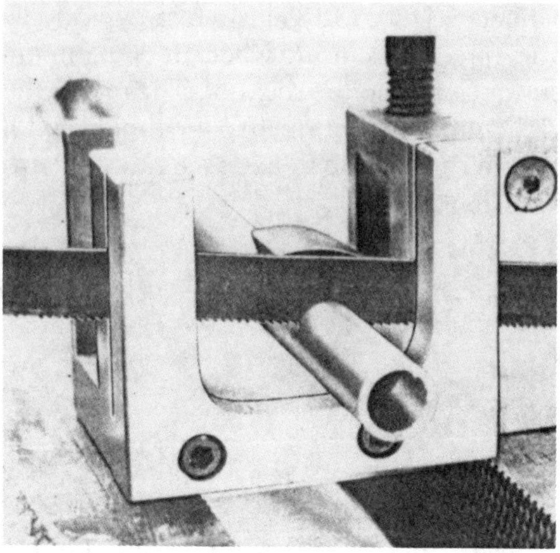

Figure 13.22 Hacksaw and guide block used to cut tubing. (Courtesy Crawford Fitting Company)

blades should have at least 32 teeth per inch. Note the soft metal jaws in the bench vise made of heavy sheet brass or copper used to avoid damage to the tubing.

Of course the cutter throws a burr into the internal diameter of the tubing (see Figure 13.23), and a hacksaw will burr both the internal diameter and the outer diameter of the tube (see Figure 13.24). These burrs should be removed after cutting the tube. Make certain that chips of metal are cleaned from the end of the tubing, as metal chips can cause a fitting to leak or cause damage to components in the system.

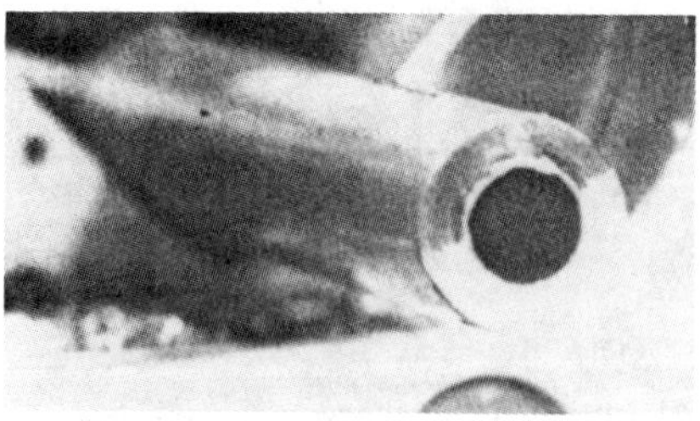

Figure 13.23 Tube cutter burrs — Inside diameter. (Courtesy Crawford Fitting Company)

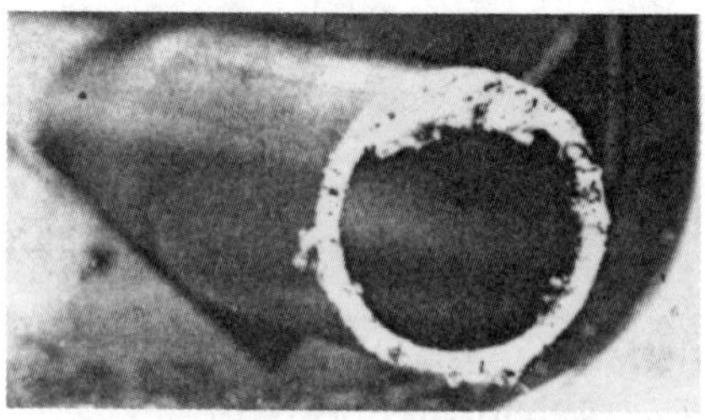

Figure 13.24 Hacksaw burrs — Inside diameter and outside diameter. (Courtesy Crawford Fitting Company)

TUBE DEBURRING

Both internal diameter and outer diameter should be deburred and cleaned after cutting, no matter which method (tube cutter or hacksaw) is used for cutting the tube. Inside diameter deburring, while not so important to the function of certain types of tube fittings, is important from the standpoint of system cleanliness. Small burrs, chips and slivers of metal can plug small pilot holes or vents, can scratch valve seats or stem tips, and can damage soft seals such as O-rings. Also the tube internal diameter can be substantially reduced, affecting the through-flow.

Outside diameter deburring is important for proper fitting functions as well as for clean, leak-free systems. Small outer diameter burrs may prevent tubing from being fully inserted through nut, ferrules, and body bore (Figure 13.25). It is vital, of course, that the tube *always* pass through the nut and ferrules and be bottomed fully against the shoulder of the fitting body.

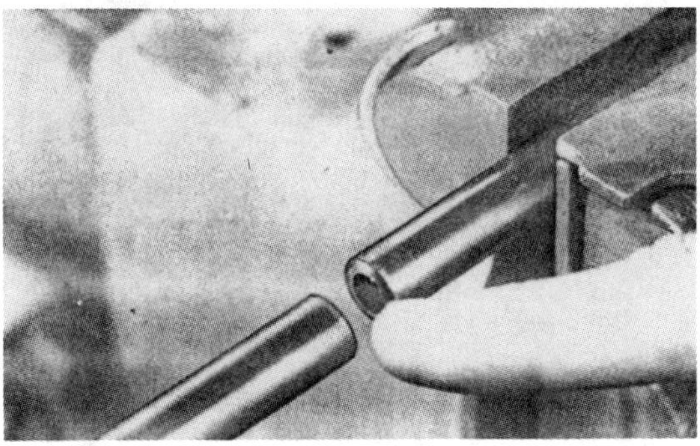

Figure 13.25 Small burrs on the outer diameter of tubing may prevent tubing from proper insertion and retention in the tube fitting. (Courtesy Crawford Fitting Company)

Even if outer diameter burrs do not interfere with full insertion, they can lodge between sealing surfaces, or break off and foul the downstream system just like an internal diameter burr. Outer diameter burrs can also scratch the fine finish of ferrule bore, making ferrule-to-tube sealing more difficult.

DEBURRING TOOLS

Outside deburring can be accomplished with a fine tooth file (Figure 13.26). Inside deburring can be accomplished using a round file, a scraper, a taper reamer or a combination design deburring tool shown in Figure 13.27.

Figure 13.26 Deburring outer diameter with a file. (Courtesy Crawford Fitting Company)

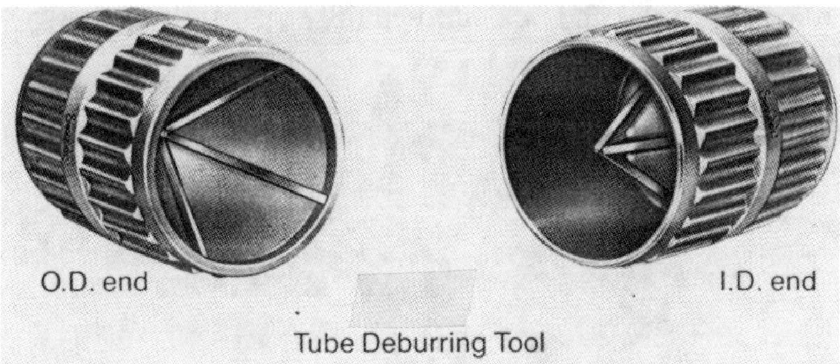

Figure 13.27 Tube deburring tools. (Courtesy Crawford Fitting Company)

THREAD PROTECTORS

Metal or plastic thread protectors should be left on any exposed pipe or tube threads until ready for immediate assembly (Figure 13.28).

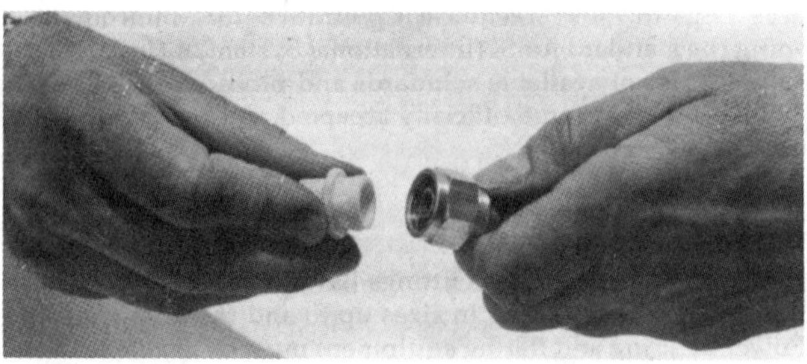

Figure 13.28 Thread protector.

TUBE INSTALLATION FOR CONVENIENT SERVICING

Many lines may be needed in very close quarters. Figure 13.29 shows a typical panel with quality bending so that all components are serviceable.

Figure 13.29 Close-up of oil well control panel showing intricate use of quality bending so that all components are serviceable. Note neatly stacked straight runs. (Courtesy Automation U.S.A.)

HOSE AND TUBING OPTIONS

Beyond the compressed air distribution system composed of rigid main pipes, feeder lines, and associated fittings and accessories, a means must be provided for conducting clean, dry compressed air and air-borne lubrication to tooling and equipment. This requirement is two-fold in that both movable and stationary tooling and equipment is involved. In addition, an almost endless variety of compressed air circuits is required for air-powered machines, instruments, gages and equipment. To fulfill these requirements, air hose and both flexible and rigid tubing are used.

Many types of "rubber" or rubber-like air hoses are available for a wide variety of applications. They are of three general types: (1) wrapped, (2) horizontal braided and (3) vertical braided. They may be electrically conductive or in some important installations, non-conductive as an insulating function where electrical devices are involved. In selecting air hose for a particular application, certain basic factors should be considered. The hose should have a smooth bore; be resistant to oil mist and vapor transmitted through the distribution system from the compressor, and also be resistant to the lubricant being carried to tools or equipment from an air line lubricator. The carcass or walls of the hose must be sufficiently strong to resist heavy impacts and shock blows, and the outer structure must be sufficiently abrasive-resistant. Since air hoses possess a wide degree of variation in their flexibility and non-kinking qualities, care must be exercised in selecting hoses which have the required characteristics. There are a number of installation techniques which should be practiced to ensure proper hose usage:

- Avoid use of excessive hose lengths.
- Avoid subjecting hose to chemicals which will destroy it.
- Provide sufficient hose length to make possible a natural loop between fixed connections.
- Do not expose hose to sunlight or excessive heat.
- Use proper fittings and couplings for all hose connections. Check sizes to avoid excessive pressure drop.
- Install fittings and couplings in a careful and proper manner.
- Make fullest use of best hose positioning techniques to provide maximum floor safety conditions, minimum hose rubbing against structures or surfaces which will cause wear and abuse, and required freedom of tool movement.

KEEP LENGTH OF AIR HOSE AS SHORT AS POSSIBLE

The length of air hose used for portable power tool applications is especially important. This is illustrated in the following example. A 50-foot length of 1/2-inch rubber air hose, having couplings at each end and through which there is a pulsating flow of 50 cubic feet per minute, will have a pressure loss of approximately 9.5 psig at a line pressure of 90 psig. With 25 feet of hose, the pressure loss through the hose will be reduced approximately 50 percent.

Hose length should be at a minimum consistent with maneuverability of the tool and the tool travel distance. Shorter lengths, in addition to having less pressure drop, are advantageous for the more rapid conveyance of oil mist from the air line lubricator to a new tool. Also, there is less probability that the oil mist will coalesce before reaching the tool.

Determination of air pressure where a "rubber" hose is used may be made with a needle-type pressure gage, which measures pressure in psig. The small portable gage, equipped with a sharp hypodermic-type needle, may be used to pierce rubber hose to obtain a pressure reading. When properly inserted and withdrawn, its infrequent use on a given hose area does not leave an air leak in the hose. The gage is particularly useful in determining the actual line pressure being delivered at the tool's air inlet. This may reveal that the tool is underpowered and that remedial measures are necessary.

Plastic tubing is also used for portable power tools and other compressed air applications. This is done in instances where the degree of flexibility and other operating conditions may permit utilization of certain advantages, such as smaller diameter and reduced internal wall friction. Coiled or self-storing (retractable) plastic tubing may be used to advantage in various instances in conjunction with portable power tools. When not in use, this tubing retracts to hang in coiled position above the floor or working level, thus holding its damage and wear to a minimum. In a similar manner, "rubber" air hose may be arranged to be retractable on a looped basis when installed in conjunction with portable power tool balancers. Both plastic tubing and "rubber" air hose may be arranged to be moved horizontally back and forth when attached in a hanging-loop manner to suspended free-moving hangers or trolleys. Recent developments of commercially-made units of overhead-traveling compressed air distribution systems are available. Also available are special compressed air ducts which permit their being tapped at various points along their length by one end of a flexible tool-connected air line.

OPTIONAL LINE MATERIALS

Other types of tubing used for compressed air applications consist of various kinds of plastic, "rubber" and metal. These are both rigid and flexible types. Examples of rigid types are copper, nylon, polyvinyl chloride and stainless steel. Flexible types include Neoprene, Polyurethane, Nylon, Polyethylene and Vinyl. Such types of tubing have in many instances replaced small diameter iron piping because of greater flexibility, lighter weight, better corrosion resistance, less frequent maintenance, smaller size, and better resistance to chemical attack. "Rubber" air hoses are also used on many types of air-operated machine components formerly equipped with iron piping.

HOSE AND TUBE FITTINGS

Permanent-type hose connections are made by using a wide variety of metal fittings of two general types, each of which is available as a rigid or swivel unit. Flared or barbed metal fittings, as shown in Figure 13.30, are

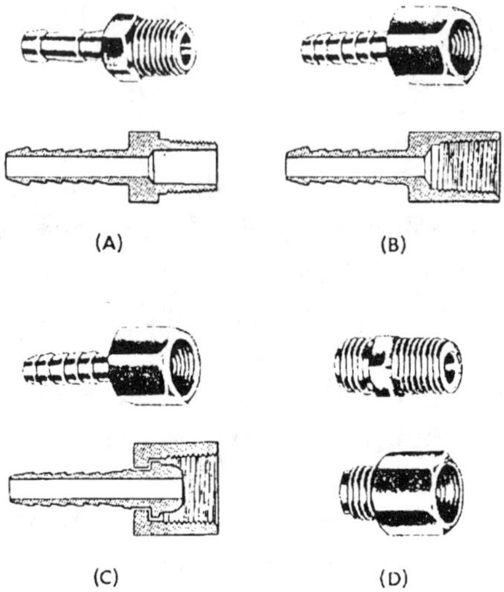

Figure 13.30 Barbed hose fittings: (A) rigid male; (B) rigid female; (C) swivel female; (D) adapters, male and female. (Courtesy Perfecting Service Company)

especially useful for conserving hose assembly time and when the entire hose assembly is worn and discarded. Ferrules which cover the end of the hose into which the barbed end of the fitting is inserted may be crimped with small crimping dies or crimping tools. Single barbed fittings having two barbed male ends are also available.

Re-usable metal hose fittings (Figure 13.31) are assembled by slipping the ferrule unit over the square-cut end of the hose, after which the stem fitting is screwed tightly into the ferrule.

In the application of fittings to hoses, more airflow (cfm) may be obtained by using the next size hose. For example, when a $^1/_4$-inch N.P.T. fitting for any pressure loss for a given length of conduit is used, a $^3/_8$-inch hose rather than a $^1/_4$-inch hose can be applied. Similarly, a $^1/_2$-inch hose may be used on a $^3/_8$-inch fitting.

Nylon tube fittings such as tees, elbows, port connectors and the like are also used for a wide variety of applications requiring tubing connections. When used with flexible plastic tubing, such fittings are self-aligning.

QUICK-DISCONNECT COUPLINGS

The versatility of many pneumatic systems may be dependent on the use of quick-connect and quick-disconnect between a tool and the pneu-

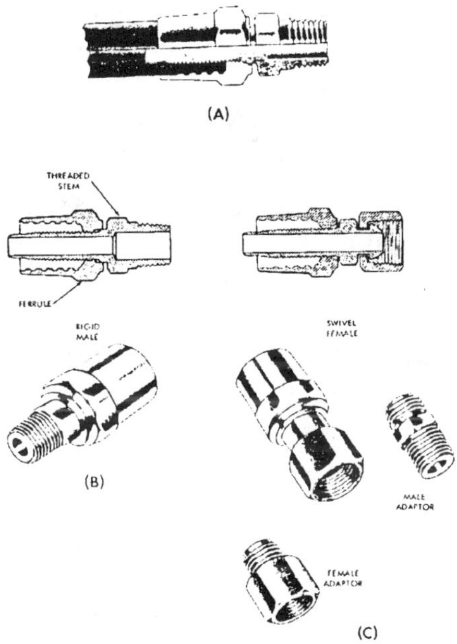

Figure 13.31 Re-usable hose fittings: (A) re-usable rigid male coupling assembled on hose; (B) assembled rigid male coupling consisting of ferrule and threaded stem; (C) swivel female coupling and male and female adapters. (Courtesy Perfecting Service Company)

matic supply. Thus, quick-disconnect couplings may be considered an essential component within a compressed air system. They are frequently used on assembly lines directly between the air hose and the portable air tool. The quick change feature allows a tool to be connected or disconnected with minimum downtime and provides for fast movement of a tool from one location to another. With reciprocating air tools it is best to have a short length of whip hose connected to the tool before making the coupling installation. This prevents the quick-disconnect coupling from receiving direct vibrations set up by the tool. Vibration causes excessive wear to the coupling and may even cause an accidental disconnect.

Quick-disconnect couplings may also be installed on the compressed air feeder lines or manifolds so that the number of individual hoses required for connecting air-operated equipment may be minimized. When equipment is not in use, the tool and air hoses can be rapidly disconnected and carried to locations where needed. Permanent-type fittings would make this impractical in many cases, since the downtime required for removal and re-fastening of the fittings to hoses would be too costly, and might require the use of more hoses.

Quick-disconnect couplings reduce the need for manual shutoff valves. The couplings feature automatic shut-off, eliminating the need for an operator to go back to some point in the line and manually close a valve. Despite their advantages, quick-disconnect couplings should be used sparingly to minimize pressure drop. Sizes of $^1/_2$-inch NPT and larger may be difficult to couple or uncouple unless the air pressure can be shut off during the process.

The time-saving feature of quick-disconnect couplings makes them invaluable where test setups are involved. With their use, the time required to "make and break" connections many times per hour is negligible when compared with the time required to "make and break" connections with screw-type fittings. This savings in time allows the quick-disconnect coupling cost to be justified in most installations.

The use of quick-disconnect couplings on any line in the air system, other than the main or header lines, is advantageous for air line maintenance because of the short time required for removal and replacement of the line.

Quick-disconnect couplings for use in compressed air systems are of three types: (1) *single shut-off*, only one half with valve (Figure 13.32); (2) *double shut-off*, with two valves (Figure 13.33); and (3) *straight through*, with no valves in either half (Figure 13.34).

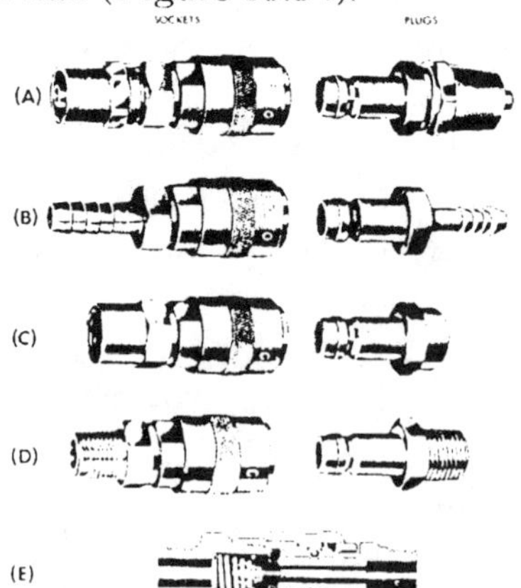

Figure 13.32 Single shut-off couplings: (A) re-usable hose connections; (B) hose stem-end connections; (C) female end connections; (D) male end connections; (E) cutaway, connected position, plug and socket, male end connections (Courtesy Perfecting Service Company)

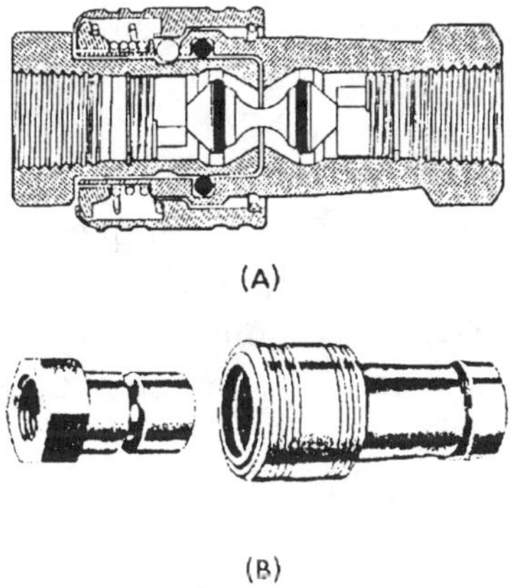

Figure 13.33 Double shut-off coupling: (A) cross section of connected plug and socket; (B) plug and socket. (Courtesy Perfecting Service Company)

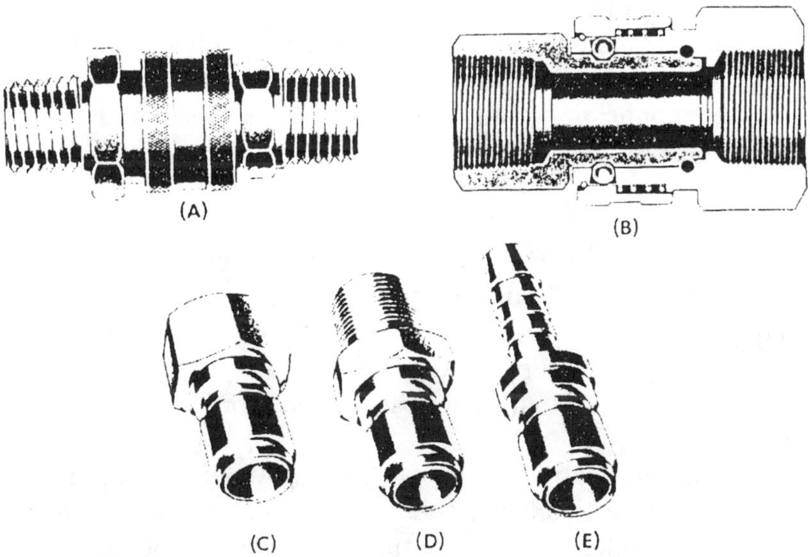

Figure 13.34 Straight-through quick-disconnect coupling: (A) connected plug and socket with male end connections; (B) cutaway, connected plug and socket, with female end connections; (C) plug, female end; (D) plug, male end; and (E) plug, hose stem end. (Courtesy Perfecting Service Company)

Single-valve couplings are used with the shut-off half installed on the air supply side of the air tool. Thus, when disconnected, the air supply automatically seals off without requiring the manual closing of a hand valve.

If it is undesirable to lose pressure on either side of the air-operated equipment when disconnecting the line, double shutoff couplings are used on lines smaller than $^1/_2$-inch NPT.

Straight-through couplings are used in locations within the air system where a quick connect or disconnect is infrequently required. There, unrestricted air flow is required, but where it may be necessary to disconnect the line, the straight-through coupling is used. A hand valve is provided in the line to shut off the air before breaking the connection.

ROTARY AND SWIVEL JOINTS

Rotary joints, rotary connectors, or rotary pressure joints are mechanical sealing devices used to conduct compressed air at a fixed angle from an air line to a revolving machine member or facility such as a rotating lathe chuck using compressed air to clamp the work piece or product. Such joints make possible the conveyance of compressed air through rotary devices such as shafts and axles. They are commonly used to actuate air clutches, air cylinders and other rotating equipment. Swivel joints make possible the connection of air lines which may be positioned and moved with rotary and angular motion, separately or simultaneously.

SUMMARY

Changes in direction of air flow creates restrictions. For most efficient movement of compressed air minimize use of elbow, "Y", "T" and return bends.

Provide adequate support for pipe, tube and hoses. If pipe lines will be used as ladders - provide adequate support for the maximum anticipated load.

Pipe and tube fabrication requires good workmanship. For minimum downtime, maintenance and zero leakage, burrs must be removed, threads must be clean, and joints must be properly tightened.

Thread protectors should be left in valves, fittings and piping until ready for installation. Spare hose should be stored with thread protector on each end. Keep air hoses as short as possible. Keep lines from valves to cylinders short to avoid captive air within the lines reciprocating with each cylinder movement, creating excessive heat and sludge formation from the captive lubricant.

Quick-connect couplings in sizes of 1/2-inch NPT and larger may be difficult to engage and disengage if the air pressure cannot be released or blocked prior to coupling or uncoupling.

REVIEW QUESTIONS

1. There are many choices of fittings for plastic tubing. What features should be compared before selection?
2. What is the purpose of pre-taping fittings?
3. What are the reasons to avoid the use of metal pipe fittings?
4. What are the major limitations of taper pipe threads?
5. What is the importance of conductive and/or non-conductive hose?
6. Why is it important to keep air hose length as short as possible?
7. What are the three types of quick-disconnect couplings typically used in pneumatic systems?
8. Name three types of plastic tubing available for use in pneumatic systems. What are the advantages and disadvantages of each type?
9. Describe a major reason for using rigid metal pipe in a pneumatic system.
10. Are shut-off valves desirable when using quick-disconnect couplings in sizes above 1/2-inch?

14

CONTROL VALVING

Various conditions of flow and pressure must be provided and controlled in the use of pneumatic power transmission systems. This is the function of valves designed for service in pneumatic systems which control the direction of flow, rate of flow, or the pressure level of the contained fluid in various parts of the power transmission circuit.

Pneumatic control valves are of three major types: (1) *pressure-control valves* (or pressure regulators) control the air pressure or variations in the pressure; (2) *directional-control valves* direct the compressed air to designated points and provide the required number of flow paths; and (3) *flow-control valves* control air flow rate. A thorough knowledge of the construction and action of pneumatic control valves is essential for design of the pneumatic circuit. The valves will be employed to provide the control functions when moving machine members by means of the pneumatically-powered force and motion actuators. The relationships resulting from the use of pneumatic power components must be carefully considered as the basic machine is designed. Safe, efficient power transmission can be provided by use of the appropriate valves and actuators for the force and motion functions. A working knowledge of the pneumatic circuit and components must be acquired by the maintenance and service personnel.

There are various ways in which the large variety of air control valves may be classified. Some differences in terminology and in the emphasis of classification criteria exist, although much progress has been made in the area of classification and terminology standardization. Several examples of differences are: (1) the term *poppet valve* may be used in one instance to designate a basic valve design and another reference may described the valve as a directional-control valve to signify its major function; (2) the term *operator* and *actuator* are used synonymously, although the latter is used by some to refer only to cylinders or motors, rather than to valves, and (3) a pressure-control valve for hydraulic fluid may be referred to as a *pressure-reducing valve*, whereas an air valve of this type is usually called a *pressure regulator.*

The Terminology Committee of the National Fluid Power Association was established in 1958. The purpose of this committee, in part, was to list technical terms in the fluid power industry and define these terms. Subsequently, this association has developed and published a "Glossary of Terms for Fluid Power." A summary of valve terminology, taken from a portion of this glossary, is shown in Table 14.1.

A complete presentation of the function, design and characteristics of the rather complex variety of air control valves is beyond the scope of this presentation. Information is provided primarily to orient the reader with and explain the most widely used types which are classified as *pressure-control, directional control,* and *flow-control valves.* A variety of circuits employing air valves is shown throughout this book with valves identified by their classification criteria.

PRESSURE CONTROL VALVES

Compressed air, when transmitted from a compressor through piping is subject to pressure and flow variations. Pressure control valves or pressure regulators are therefore required to (1) reduce the pressure when it is too high for the equipment or operation involved; (2) provide relatively constant pressure to equipment when the line pressure varies, and (3) stabilize inlet pressure to the actuators when the air flow varies.

Pneumatic pressure-control valves or pressure regulators are often poppet-type valves positioned by diaphragms or pistons. Cutaway views of some widely used types and a description of the operation may help understand the function in the pneumatic circuit.

Section 72 Valves–Functional Types	Section 73 Valves–Basic Design	Section 72 Valve Operators and Positions
Air	Jet Action	Operator (Valve)
Directional Control	Flapper	Manual
Check	Seating Action	Mechanical
Diversion	Ball	Pilot
(see Selector Valve)	Diaphragm	Barrier
Four-way	Disc	Differential area
Selector	Swing	Differential pressure
Straightway	Gate	External
Three-way	Spreader	Internal
Flow Control	Wedge	Solenoid controlled
Deceleration	Globe (see Disc Seating)	Solenoid
Pressure compensated	Needle	Position
Pressure-temperature compensated	Plug	Center
Flow Dividing	Poppet	Closed
Pressure compensated	Shear Action	Open
Flow Metering	Plug	Center pressure
Hydraulic	Plunger	Detent
Pilot	Spool	Float
Pneumatic	Sliding plate	Hold
Prefill	Linear	Normal
Pressure Control	Rotary	Normally closed
Counterbalance		Normally open
Decompression		Offset
Load-dividing		Regenerative
Pressure reducing		Return
Relief		Four position
Safety		Three position
Unloading		Two position
Priority		
Sequence		
Shutoff		
Shuttle		
Surge damping		
Time delay		

Table 14.1 Summary of Terminology Relating to Valves. (Glossary of Terms for Fluid Power - National Fluid Power Association, 3rd interim edition, November, 1962).

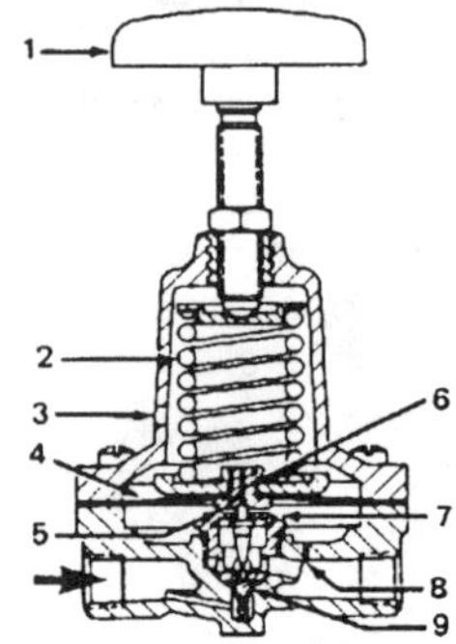

Figure 14.1 Diaphragm-type pneumatic pressure regulator. (Courtesy C. A. Norgren Company)

In the diaphragm-type valve shown in Figure 14.1, downstream pressure (the controlled function) is allowed to act on one side of the diaphragm, the movement of which is generally opposed by a spring or reference pressure on the other side. An increase of downstream pressure causes movement of the diaphragm in a direction to reduce the flow area at the closure element. A decrease of downstream pressure produces movement to open the closure element and increase the flow area.

The working elements of this pressure regulator are the adjusting knob (1) or slotted adjusting screw, regulating spring (2), diaphragm (4), valve seat and pin assembly (7), and valve ball (9). The lower side of the diaphragm is referenced to the outlet (secondary) side of the regulator through aspirator passage (8).

With the adjustment turned counterclockwise until no load is applied to the regulating spring, the valve ball is held in the closed position by the valve spring and inlet pressure. When the adjustment is turned clockwise, a force is applied to the regulating spring (2) which causes the diaphragm to move downward. This motion is transmitted through the valve pin, unseating ball (9) and allowing air to flow downstream. The increase in outlet (secondary) pressure, sensed under the diaphragm through passage (8), creates a counteracting force tending to move the diaphragm upward, compressing the regulating spring and allowing the valve ball to begin moving upward towards its seat. This upward movement will continue until the pressure force exerted on the lower side of the diaphragm balances the spring force exerted on the upper side. If there is no downstream flow demand, this balance of force occurs just before the valve ball contacts its seat. If there is a flow demand, the balance of forces will occur with the valve ball held further off its seat just the amount necessary to compensate for the demand, thus maintaining the desired downstream pressure.

This regulator has a constant bleed feature which helps to minimize outlet pressure drop when a flow demand is initially placed on the regulator. A very small amount of secondary air flows continuously through bleed passage (5) in the diaphragm and escapes to atmosphere through vent (3). This keeps the valve slightly open to replace the air lost to atmosphere through the constant bleed. Since the valve is always partially open, the pressure drop is minimized when demand is initially increased from no flow to some higher flow. This constant escape of air from the regulator bonnet is normal and does not indicate a faulty regulator.

This regulator is a relieving-type which allows reductions in secondary pressure under dead-end conditions by automatically relieving over pressure in the downstream line. Should a secondary over-pressure occur, whether as a result of a reduction in regulator pressure setting or some other cause, the diaphragm will move further upward and open relief passage (6) in the diaphragm. This allows secondary air to escape into the regulator bonnet and then to atmosphere through vent (3). Depending upon the source of the original over-pressure, the secondary (outlet) pressure will reach a level between the initial setting and the primary (inlet) pressure to the regulator. If the over-pressure condition is completely relieved, the diaphragm will move downward and close the relief passage.

Turn the adjustment (knob or slotted screw) clockwise to increase secondary pressure, counterclockwise to decrease the pressure. After the desired pressure setting is completed, tighten jam nut to lock pressure setting.

PILOT-OPERATED REGULATOR

These regulators are recommended for use in compressed air systems which require precision pressure control at an inaccessible location remote from the pilot regulator. Graphic symbols are shown in Figure 14.2*a* with and without a pilot. The physical size of the piloted unit can be any size appropriate to the circuit needs (Figure 14.2*b*). Several features can be gained with the remote pilot function: a) Design flexibility allows main regulator to be installed at any point in the compressed air system without regard to accessibility. Pilot regulator may be installed at most convenient location. b) Large air passages and valve openings permit high flow capacity. c) Accurate pressure regulation over a wide flow range. d) Self-relieving feature permits the regulated pressure to be adjusted under dead-end conditions. e) Constant bleed helps ensure high accuracy of regulation.

Operation (Figure 14.2*c*) - Rotation of the pilot regulator's adjustment screw in a clockwise direction introduces pressure into the pilot chamber (1) which causes the main valve (2) to open, introducing downstream pressure. When the pressure on the underside of the diaphragm (3) is approximately equal to that acting above the diaphragm, the valve will close and the desired output pressure will be achieved. For accurate secondary pressure recording, the gauge should be installed on the regulator. The constant bleed flow path is through a small orifice from the underside of the diaphragm to atmosphere by way of the relief port in the diaphragm spacer.

DIGITAL AIR PRESSURE CONTROL WITH A SOLENOID PILOT ASSEMBLY/REGULATOR

The solenoid pilot assemblies of Figure 14.2*d* offers a means of electrically controlling air-piloted pressure regulators to provide regulated output pressures up to 140 psig (9.7 bar) and air flows to 1,600 scfm. These pilots are intended for use with programmable controllers, computers, sequencers or other solid state circuitry having a digital (on/off) output.

The pilot has two normally-closed 2-way solenoid valves on a common body. One of the solenoid valves admits supply air to the regulator bonnet when energized. The other exhausts air from the regulator bonnet when energized. The amount the pressure increases or decreases in the regulator bonnet depends upon the length of time the solenoids are energized (the pulse width) and the number of pulses.

The circuit shown in Figure 14.2*e* is only a representative guide to the kind of controls that can be combined with a solenoid-operated pilot regulator. Although the circuit represented here uses a controller, some applications may only require set point limit switches and solenoid pulse width timer logic.

The externally-controlled type valve shown in Figure 14.3 requires the use of a pilot pressure regulator downstream. This functions to sense the line pressure and transmit a pilot pressure to the valve, effecting a two-stage regulation.

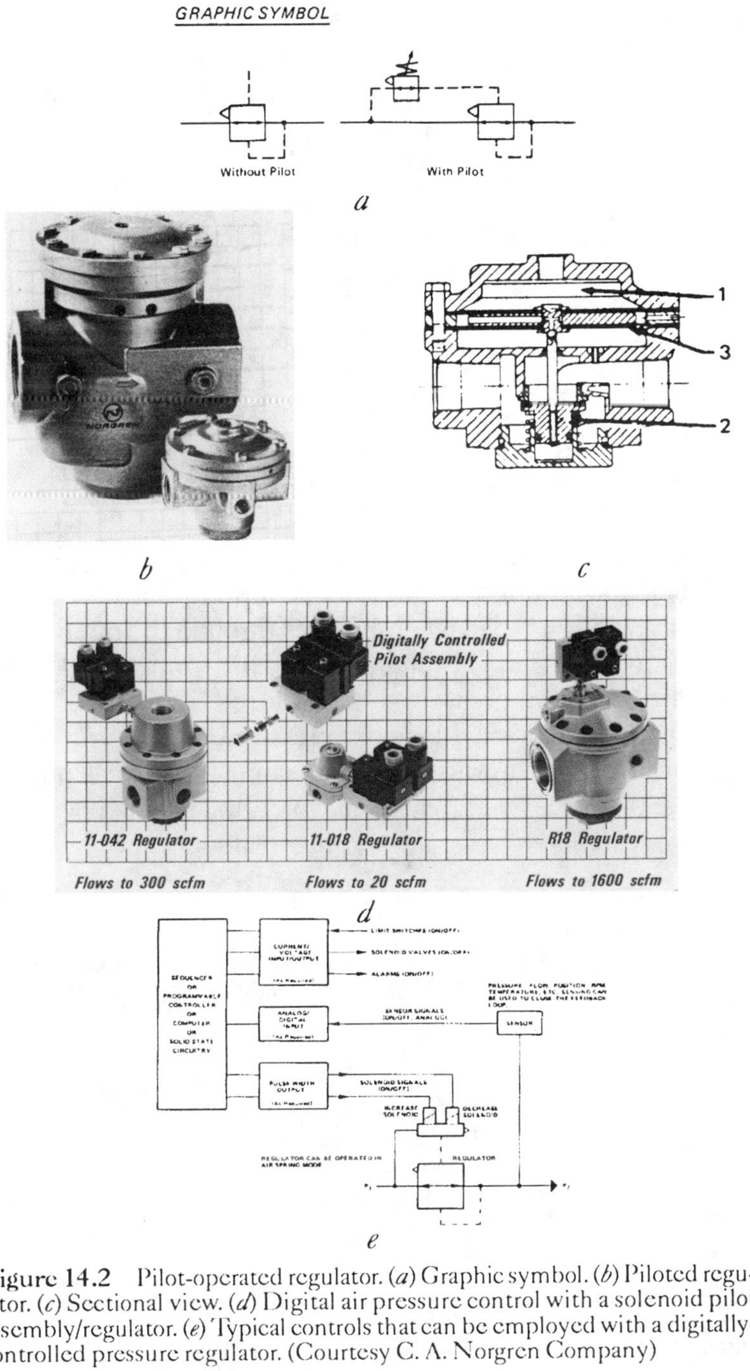

Figure 14.2 Pilot-operated regulator. (*a*) Graphic symbol. (*b*) Piloted regulator. (*c*) Sectional view. (*d*) Digital air pressure control with a solenoid pilot assembly/regulator. (*e*) Typical controls that can be employed with a digitally-controlled pressure regulator. (Courtesy C. A. Norgren Company)

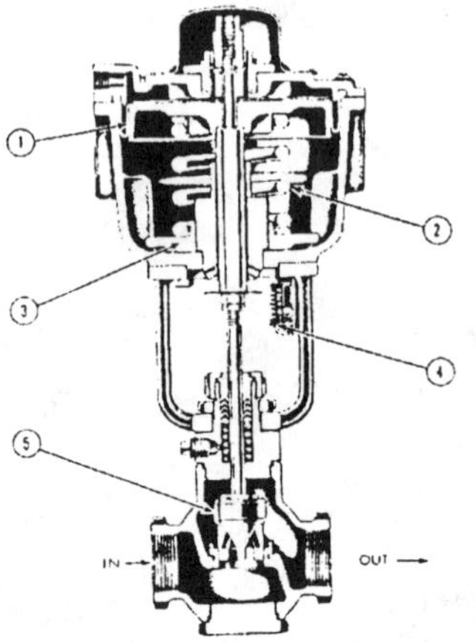

Figure 14.3 Externally-controlled regulator: (1) diaphragm, (2) spring, (3) adjusting ring, (4) travel indicator, (5) closure. (Courtesy Fisher Governor Company)

The regulator of Figure 14.4 is an example of a piston-type regulator. Details of its functioning are as follows. Assuming that air is being used downstream, it enters space A and passes through the open valve into space B. Flow from B to C is unobstructed. Pressure from C, which is downstream regulated pressure, is transmitted through port (3) to the large piston (4). If the force due to this pressure on the piston is less than spring force, the piston moves forward, opening the valve. Conversely, if the pressure produces more than spring force, the piston moves back, al-

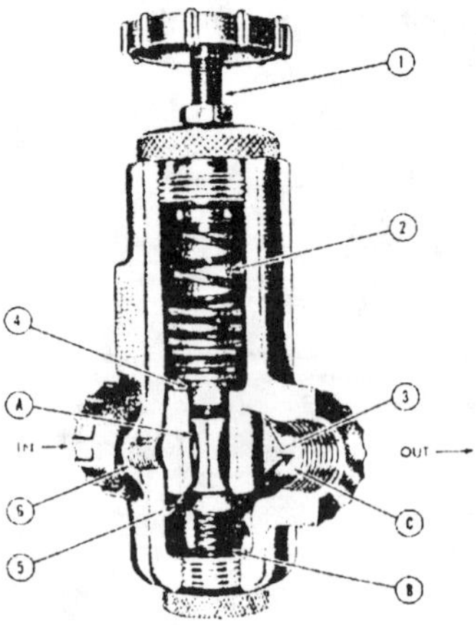

Figure 14.4 Piston-type regulator: (1) adjusting screw, (2) spring, (3) sensing pressure port (not visible), (4) piston, (5) closure, (6) pressure gage mounting hole. (Courtesy Perfecting Service Company)

lowing the valve to close. Air will flow through relief port (3) only if downstream pressure continues to rise after the main valve is closed due to an external force on the actuator within the pneumatic circuit.

OPERATING CHARACTERISTICS

The quality most desired in a pressure regulator is its capability to maintain the set pressure, even though the flow rate changes. Regulation curves or tabular data are published by the manufacturers with the set pressures at zero flow as the starting point. As flow increases, the regulated pressure generally drops, fast at first until the valve is unseated, then more slowly over a wide range, then faster again as the flow area becomes restrictive.

Some manufacturers of pressure regulators utilize venturi sections to develop a modified sensing pressure, which will aid the regulation over a portion of the pressure-flow curve. Also, to avoid the initial fast drop-off at low flow rates, a small, continuous bleed, either external or incorporated in the regulator is used. This bleed improves regulation when the regulator must control a static, or no-flow condition. Dynamic systems, or systems which have appreciable continuous flow, are more easily controlled.

Another feature desired in a regulator is its capability of maintaining set pressure with wide variations of upstream pressure. One means of avoiding this effect is to provide in the design a balancing piston of approximately the same area as the sealing area of the valve. Thus, a pressure difference across the valve is equalized by a pressure difference across the piston. Another way to minimize the effect is to provide a large control diaphragm area in relation to the valve area so that the ratio of regulated pressure change to upstream pressure change is roughly proportional to the ratio of valve to diaphragm area.

CHOICE OF REGULATOR

The choice of a diaphragm- or piston-type pressure regulator depends upon several factors such as cost, reliability and regulation capability. Piston-type units are usually considered more reliable and more easily repaired, but they may be more costly. Reliability of the piston unit is enhanced by relatively easy replacement of standard O-ring seals. The regulation of piston types, especially small one-body self-enclosed units, is affected by friction on the moving members. Diaphragm types are at a disadvantage due to the change of effective area which occurs when a large movement of the valve is required. Thus the piston types generally have larger initial pressure drop off, while diaphragm types have more pressure drop off in the middle of the range. A rolling diaphragm can overcome this fault, but at higher cost. At little extra cost, regulators are also available which will relieve the downstream pressure should it rise above the initial setting. The relief point is often not very consistent or close to the set pressure, but this feature may be important where a cylinder is being forced-closed against the supplied air. The relief point also acts as a safety valve in case the regulator main valve fails to seal completely. For complete safety, a regulator with a large relief flow capability or a separate relief valve should be installed where conditions are such that a failure of the regulator in allowing down-stream pressure to equal upstream pressure could cause extensive damage or danger to lives. Non-relieving types without constant bleed should be specified where loss of the compressed air would create a nuisance or economic concern.

DIRECTIONAL CONTROL VALVES

A directional-control valve is designed to direct or prevent flow through selected passages. This type of valve can be used to advance and retract cylinder pistons, actuate rotary air motors in a desired pattern and provide a sequence of functions as established by the fluid power circuits. The most commonly used types of directional-control valves are spool, poppet and disk, involving both seating-action and shear-or sliding-action basic design.

SHEAR-ACTION DESIGN

Shear-action or sliding-flow restriction is accomplished by incorporating an element in a valve which slides across the flow path to effect a modulating air flow. Various types of sliding members are employed such as plug, plunger, disc, piston, spool, or sliding plate. In action, such members perform in the manner of a pair of shear blades. The air flows between a pair of sharp edges, and the width of the opening can be varied from zero to a maximum.

PISTON OR SPOOL VALVE

The use of a spool, or piston sliding member is a common design feature of a shear action valve. This type of valve may be designed with various sealing means: cup packing, U-packing, O-rings or metal seals. The piston is of balanced design, to avoid unbalanced pressure loads on it which tend to increase friction and force it against the operating device. These valves are usually applied in circuits as master control valves. They may be operated either manually or they are applicable to many types of remote or pilot-operated needs. A typical spool-type 3-way valve is illustrated in Figures 14.5. Figure 14.6 shows an example of a piston-type valve with a lever operator and two position detents.

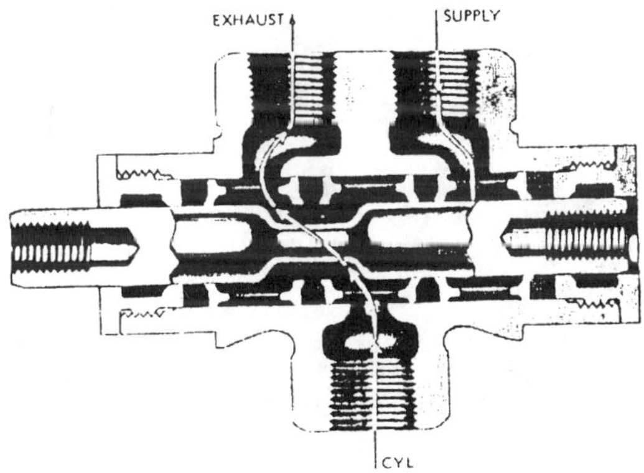

Figure 14.5 3-way spool-type valve, in exhaust position (operators not shown). (Courtesy Bellows-Valvair Division Parker Hannifin Corporation)

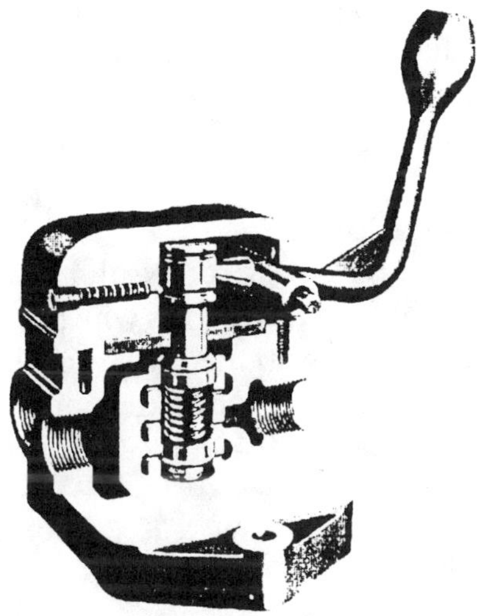

Figure 14.6 Piston-type, 4-way, lever-operated, two-detent valve. (Courtesy Logansport Machine Company)

PLUNGER VALVE

Another type of valve similar in construction to the piston type is the plunger-type directional-control valve, shown in Figure 14.7. Flow is from the inlet chamber through the radial ports into the hollow plunger and out through other ports in the plunger to the connected line.

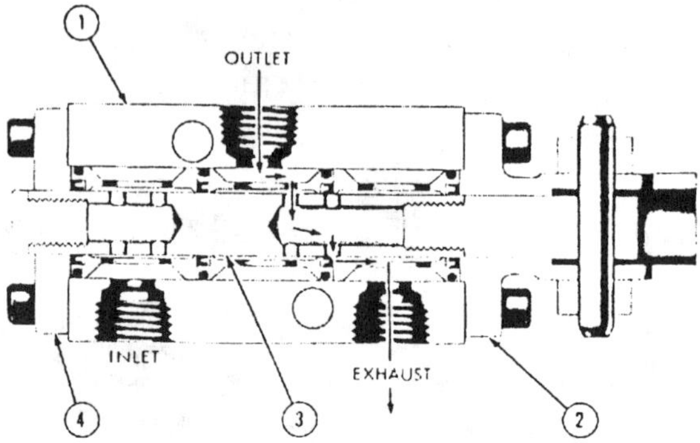

Figure 14.7 Plunger-type, 3-way directional-control valve in exhaust position (operators not shown): (1) housing, (2) handle cap, (3) plunger, and (4) end cap. (Courtesy Bellows-Valvair Division Parker Hannifin Corporation)

DISK VALVE

The disk-type valve is ideal for low pressure and high temperature applications because it needs no packing. Typical construction is shown in Figure 14.8. In the valve illustrated, the movement of the disc is a combination of rotary and lateral sliding on mating surfaces; it is mounted on a cylinder or where it can be adequately supported on the piping.

PLUG VALVE

One of the oldest type valves is the rotary-plug valve which has a flow passage or a system of drilled or cored holes which coincide with passages in the body when the plug is positioned (Figure 14.9). The air pressure would raise the tapered plug, but spring tension keeps it in place. It is used for manual control of single-or double-acting air cylinders and also as a pilot valve.

SEATING-ACTION DESIGN

Seating action is obtained by incorporating a seated obstruction in the flow path to stop the air flow. Types of seating-action valves include ball, disc, gate, globe, needle, poppet and diaphragm.

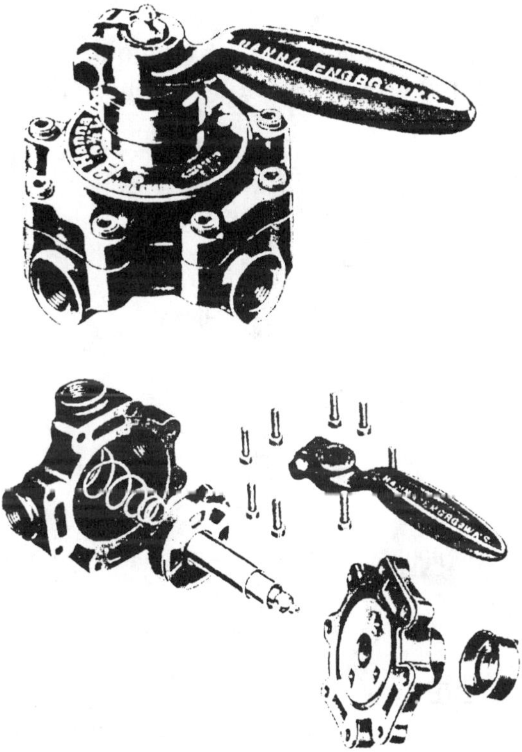

Figure 14.8 External and exploded views showing typical construction of a disc-type, 4-way, lever-operated, 3-position valve. (Courtesy Hanna Company)

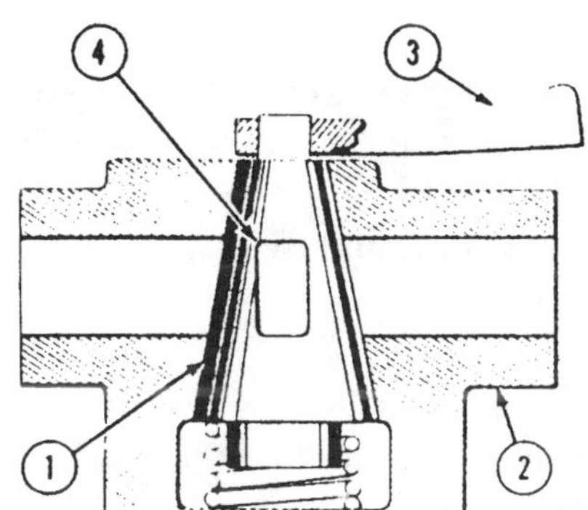

Figure 14.9 Cross-section of a lever-operated, 2-way, 2-position, rotary plug valve: (1) plug, (2) body, (3) handle, (4) flow passage.

POPPET VALVE

Poppet-type valves permit rapid opening, but require only a small travel distance for full-volume flow (Figure 14.10). The valve seats can be of flat disc or washer design.

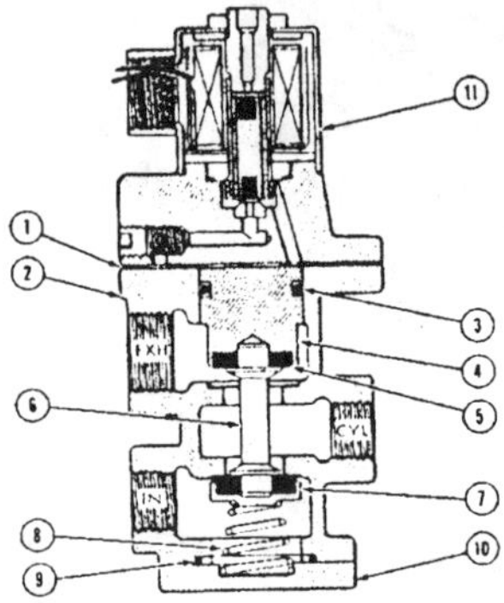

Figure 14.10 3-way poppet valve, solenoid pilot-operated, with spring return, in normally-closed position: (1) gasket, (2) body, (3) O-ring, (4) piston, (5) seal, (6) push-rod, (7) poppet, (8) spring, (9) O-ring, (10) cap, (11) solenoid pilot assembly. (Courtesy C. A. Norgren Company)

DIAPHRAGM VALVE

A diaphragm-type directional-control valve utilizes a diaphragm to obstruct the flow path. A diaphragm of synthetic materials controls the opening and closing with changes in pressure. A variety of materials is used in diaphragms made by various manufacturers.

VALVE PORTING

The number of connections made to the body of a directional control valve is governed by the type of cylinder, motor or circuit to be controlled. The valve flow-path arrangements are typically classified as two-way, three-way, four-way and five-way.

TWO-WAY VALVES

Two-way valves such as the plug cock of Figure 14.9 have two ports which provide for flow in either of two directions. In one position, the outlet is closed to pressure and in the other, free flow of pressure to the outlet is provided. Typical two-way valves differ from three-way valves in that the two-way valves do not provide for discharge of downstream air.

THREE-WAY VALVES

Three-way valves have a minimum of three ports: inlet, outlet and exhaust (Figure 14.5, 14.7, 14.10). In one offset position, the outlet is closed to pressure and open to exhaust. In the other offset position, the outlet is open to pressure and closed to exhaust. This action can be modified so that in the center position, the outlet is blocked to both pressure and exhaust, thereby giving a holding action to the cylinder (Figure 14.11). The main use of this type valve is for applications where the cylinder or ram does not require air pressure to return it to a rest position. By plugging the exhaust port, some three-way valves can be converted to two-way valves.

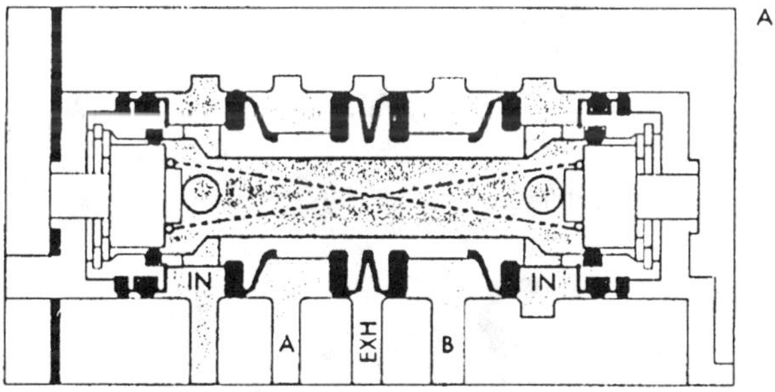

Figure 14.11 3-position centering. Three-position solenoid and remote air pilot valves can be centered by a patented spring centering device or patented combination spring and pressure-assisted spool design which reduces side load potential and resultant wear, and assures fast, positive return of the main spool when the pilots are de-energized. (Courtesy MAC Valves, Incorporated)

FOUR-WAY VALVES

Four-way valves have four or more ports: one inlet, two outlets (usually used as cylinder ports), and one common exhaust (Figure 14.6). Four-way valves with five ports (pressure, two cylinder, two exhaust) are also available. The main use of four-way valves is to actuate cylinders under power in both directions or a rotary air motor in two directions of rotation.

FIVE-WAY VALVES

Some four-way valves having five ports are capable of utilizing the normal supply port as a common exhaust port, and thus may be used as dual pressure five-way valves. Two inlets on five-way valves permit the use of different pressures for each cylinder stroke.

VALVE OPERATORS

A valve operator is the valve component through which force is applied to move or position the flow-directing elements of valves. Operators serve two major functions: to actuate a valve or return it to its normal or first position after it has been actuated.

There are various methods of operating the wide variety of control valves. If pressures are small, various types of manual methods are available. If a grinding table is to be cycled, for example, a mechanical cam-type operator may be used. If a remote operation is needed, a solenoid or pilot method of valve actuation may be selected.

MANUAL OPERATORS

The term *manual operator* applies to valves that are operated by hand or foot (Figures 14.6, 14.8, 14.9). Lever-operated valves with detents, spring, and air returns are used where manual positioning is desired (Figures 14.12). Button or palm button operators are used for hand operation,

Figure 14.12 Manual valve (hand) operator included in sub plate mounted valve assembly. (Courtesy MAC Valves, Incorporated)

and some are also adaptable to foot operation (Figure 14.13). They are available with detents and spring returns, and are used most widely in pairs with a spring return or air return in safety circuits, requiring two-

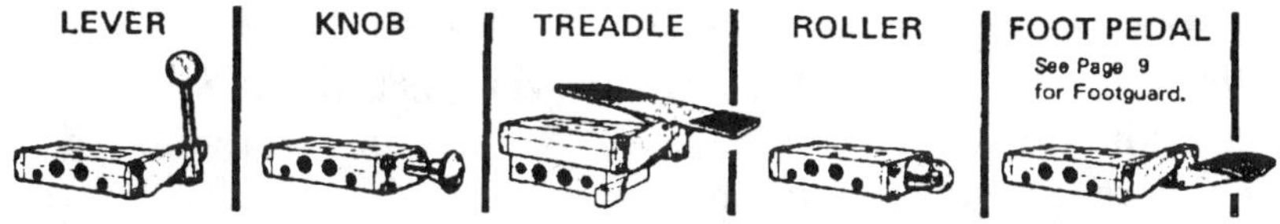

Figure 14.13 Manual operators can include lever, knob, treadle, roller or foot pedal. (Courtesy C. A. Norgren Company)

hand operation. Pedal operators require only a foot or toe movement and require a return operator such as spring or air return (Figure 14.14A, 14.14B). Treadle operators require the use of heel and toe and they usually require a spring or detent (Figure 14.14C).

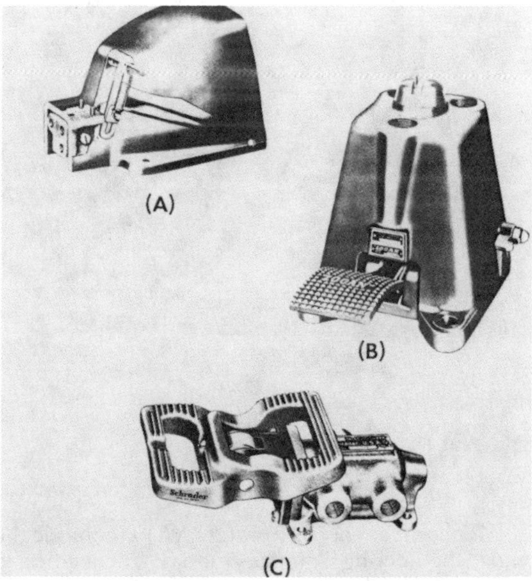

Figure 14.14 Manual valve (foot) operators: (A) with foot guard; (Courtesy A. Schrader's Son) (B) with external pedal; (Courtesy Logansport Machine Company) and (C) with lock-down rocker-type treadle (Courtesy A. Schrader's Son)

MECHANICAL OPERATORS

One of the most important applications of pneumatic controls is accomplished through a number of different mechanical means. Some of the most widely used methods are cam roller, link-clevis and mechanical lever.

The cam operator provides a mechanical method of actuating the valve by applying a variable force for various types of spring return valves. The cam operator is actuated by the force and motion of a machine element. Such operators may incorporate an override to avoid accidental damage if excessive force is applied.

A link-clevis operator provides a means of mechanical action through linkage arrangements (Figure 14.15A). A mechanical lever operator is actuated by methods similar to the cam operator (Figure 14.15B).

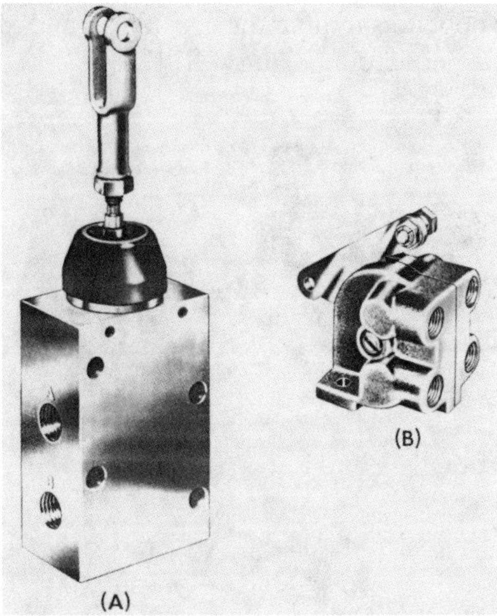

Figure 14.15 Mechanical valve operators: (A) mechanical link device (Courtesy Hanna Engineering Company) and (B) mechanical lever. (Courtesy A. Schrader's Son)

Valves with mechanical operators are ideal for many semi-automatic and fully-automatic controls. Such valves serve as inter-locks on complicated machine circuits where motions must be positive. These valves may be operated from a machine member such as a moving piston rod or a rotary-cam plate operated by a driving mechanism. Mechanically-operated valves can also be used for work-ejector service on punch presses, to blow scale from forging dies, and for applications on machine tools and other equipment where it is necessary to open or close an air line for timed intervals.

SOLENOID OPERATORS

A solenoid operator is an electromagnet containing a plunger or core, arranged to operate a valve when the coil is energized. It is used singly with a spring return, air return, or pilot return. Two solenoids are used where an electrical impulse is desirable to return the valve to its initial position. Solenoids are available in a number of voltage and current characteristics (115 volt, 60 hertz alternating current; 115 volt direct current; 220 volt, 60 hertz alternating current; 220 volt direct current; 24 volt direct current, etc.). Various types of coils are available such as varnish-impregnated coils, epoxy resin molded coils and other types designed to resist vibration, mechanical shock, moisture, heat, chemicals, etc. Solenoid enclosures of various types are also available with 1/2-inch NPT conduit connections, grommet outlets, plug-in terminals, and construction to JIC (Joint Industry Conference) standards - designed for various purposes such as being explosive-proof, water-tight, etc.

Solenoids are used to actuate poppet-type, piston-type, spool-type, plunger-type and other types of control valves. The solenoid may function as direct acting, causing a valve spool to move in a desired pattern (Figure 14.16, 14.17). The solenoid may also be used to actuate a relatively small pilot valve which in turn will pilot a larger valve (Figure

Figure 14.16 Lapped spool and sleeve design with direct solenoid actuation. (Courtesy Numatics, Incorporated)

14.18). A direct acting (high force) solenoid is one which is directly connected to the valve, directly opening or closing the valve ports, depending on whether the solenoid is energized or de-energized (Figure 14.16, 14.17). Operation is not dependent on the air pressure or rate of flow. Figure 14.19 illustrates a small three-way valve with a balanced

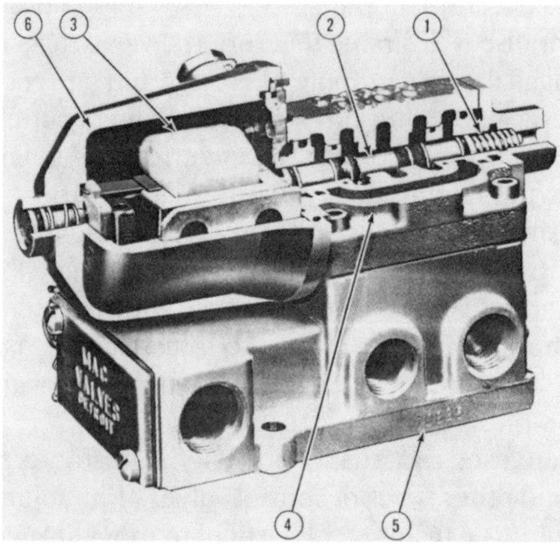

Figure 14.17 Cutaway view of a 4-way, single-solenoid, direct-operated, spring return valve with resilient seals: (1) spring, (2) spool assembly, (3) solenoid, (4) body, (5) sub-base, (6) solenoid cover. (Courtesy MAC Valves, Incorporated)

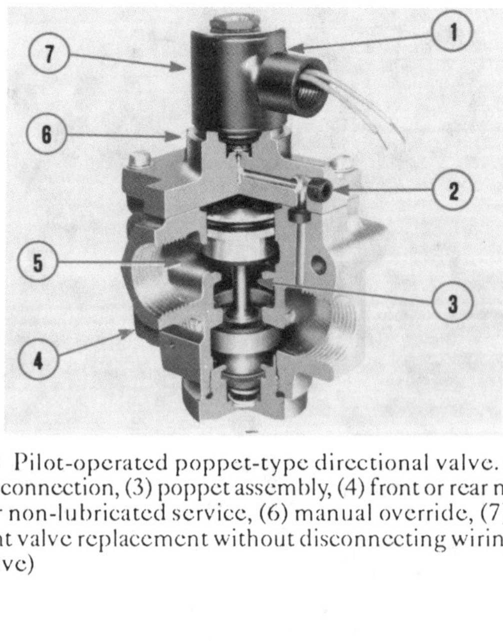

Figure 14.18 Pilot-operated poppet-type directional valve. (1) pilot, (2) External pilot connection, (3) poppet assembly, (4) front or rear mounting, (5) Lubricated or non-lubricated service, (6) manual override, (7) slip-on coil permits instant valve replacement without disconnecting wiring. (Courtesy Automatic Valve)

poppet assembly. When assembled to a 4- or 5-way valve the pilot becomes an integral part of the valve assembly (Figure 14.20)

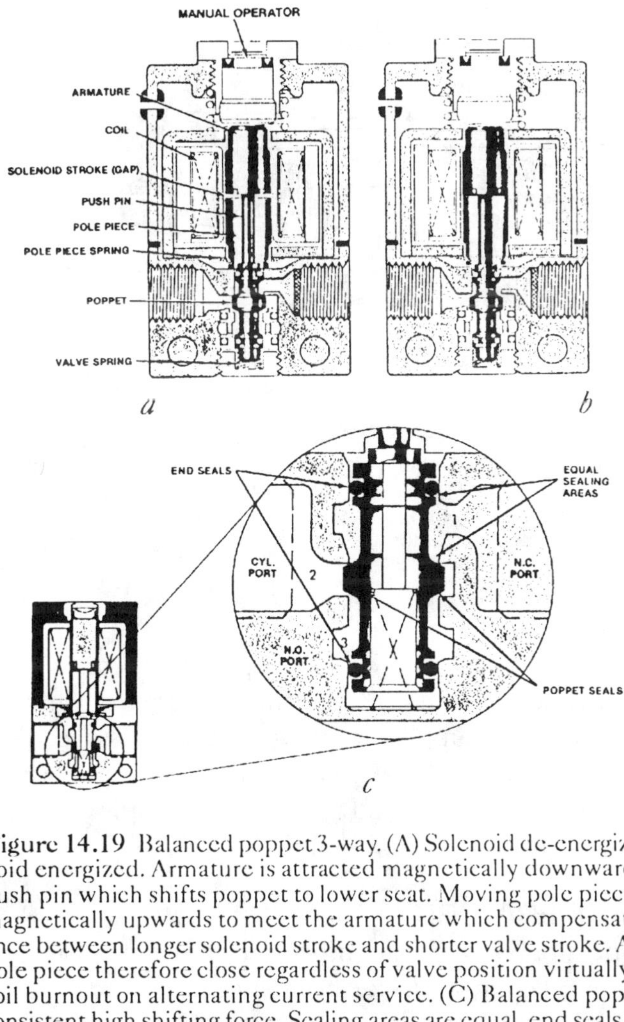

Figure 14.19 Balanced poppet 3-way. (A) Solenoid de-energized. (B) Solenoid energized. Armature is attracted magnetically downward, extending push pin which shifts poppet to lower seat. Moving pole piece is attracted magnetically upwards to meet the armature which compensates for difference between longer solenoid stroke and shorter valve stroke. Armature and pole piece therefore close regardless of valve position virtually eliminating coil burnout on alternating current service. (C) Balanced poppet provides consistent high shifting force. Sealing areas are equal, end seals balance poppet seals. Forces created by air pressure are equal in both directions. Valve can be connected for normally- open or normally-closed functions. (Courtesy MAC Valves, Incorporated)

A pilot solenoid is essentially a small, complete, direct-acting, solenoid valve which controls a pilot air pressure, which in turn actuates a piston or diaphragm that mechanically opens or closes the valve (Figure 14.20). The pilot air may be obtained through an internal passage in the valve, or it may be supplied from an external source, depending upon the valve design and requirements.

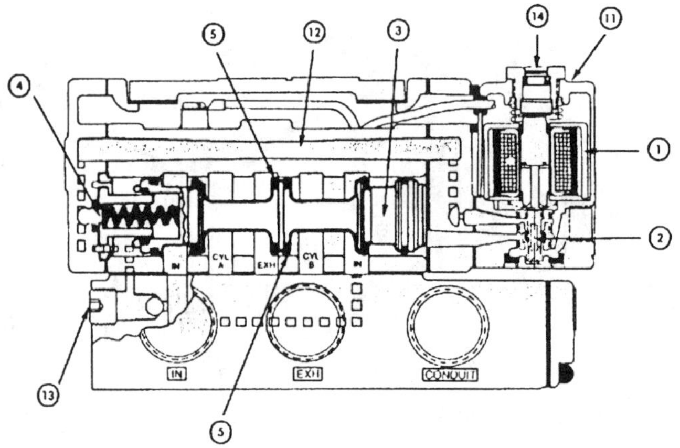

Figure 14.20 2-position, 4-way, single solenoid, internally-piloted, spool-type valve; (1) solenoid, 2) balanced poppet pilot valve (3) large spool, (4) air spring, (5) seal assembly, (11) NEMA 4-solenoid enclosure, (12) integral accumulator, (13) internal or external pilot operation, (14) manual operator. (Courtesy MAC Valves, Incorporated)

PILOT OPERATORS

Pilot operators actuate an attached valve directly by means of a piston within a cylinder. The pilots may receive air pressure from an external source or from the main air supply through an internal passage. They are dependent upon the area of the piston and the air pressure as to their function. Pilots are extensively used in the more complex circuits and circuits requiring positive interlocking and sequencing.

DIRECT-PILOT OPERATORS

A direct-pilot operator is a single-acting device similar to an air cylinder into which air is supplied from an external source to actuate the attached valve (Figures 14.21, 14.22). A pilot operator is used in conjunction with spring returns, air returns, and in some cases with mechanical returns.

LOW-PRESSURE PILOT OPERATORS

A low-pressure pilot is functionally the same as a direct pilot. It is so designated because it has a large piston or diaphragm area which reduces the pressure of the pilot air required to actuate the attached valve. The low-pressure pilot operator is used in conjunction with spring returns, air returns, and in some cases with mechanical returns.

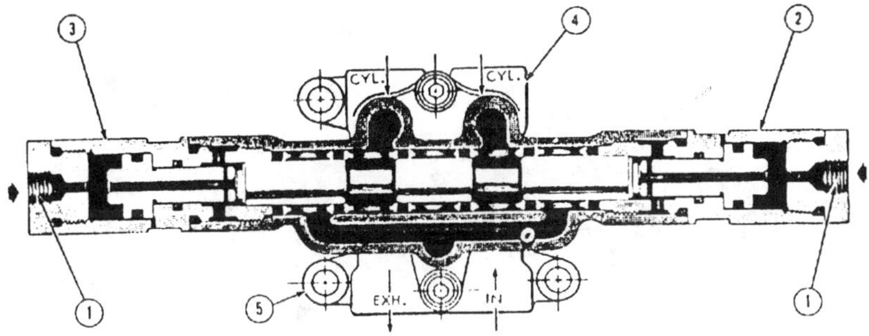

Figure 14.21 Top view of a sub-base mounted, 2-position, 4-way, 4-port, spool-type valve with double pilot operators: (1) pilot pressure inlet, (2) (3) pilot operators, (4) sub-base, (5) mounting hole. (Courtesy Bellows-Valvair Division Parker Hannifin Corporation)

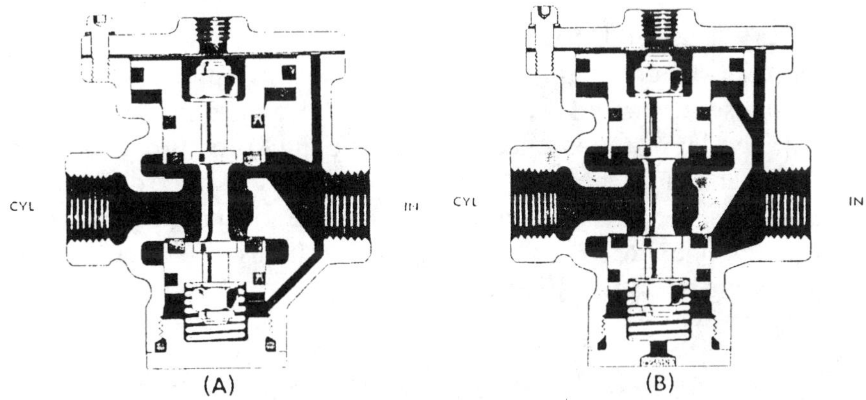

Figure 14.22 2-way, pilot-operated, poppet-type valve with combination spring-and-differential pilot return. (A) normally-open position; (B) normally-closed position. (Courtesy Bellow-Valvair Division Parker Hannifin Corporation)

BLEED-PILOT OPERATORS

Bleed-pilot operators are used on both ends of a valve, being pressurized at all times through a small internal pilot pressure passage in the valve. Relieving the pressure from either pilot chamber will disturb the pressure balance and cause the valve to shift toward the bled or lower pressure side (Figure 14.23). Bleed-pilot operators are sometimes used in combination with differential pilots to produce an automatic return. In the valve illustrated, line pressure is admitted to both ends of the spool through the cored body (1). Since the areas at both ends are equal, the thrusts are balanced. Bleeding the air from either chamber (2) upsets the balance, causing the spool to shift toward the bled or lower pressure end.

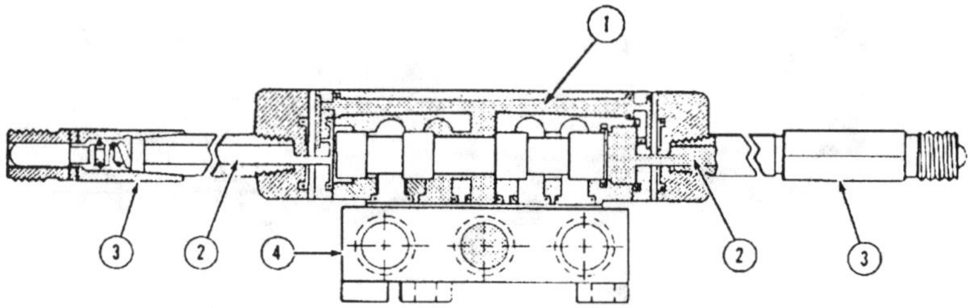

Figure 14.23 4-way, 5-port, double bleed valve, with auxiliary bleeders: (1) cored valve body, (2) bleeder operator chamber, (3) bleeder operator with ball-nosed plunger, (4) sub-base. (Courtesy Alkon Products Corporation)

DIFFERENTIAL- PILOT OPERATORS

A differential-pilot operator is basically the same as a pilot operator except that it is pressurized at all times from a pilot air source. When the differential pilot piston is acted upon by the pilot pressure, it exerts a force less than that of the opposing operator. In this way, the differential pilot acts the same as a spring return and has the advantage of faster response and longer life. A differential pilot may be used as a return with most actuating operators (Figures 14.24).

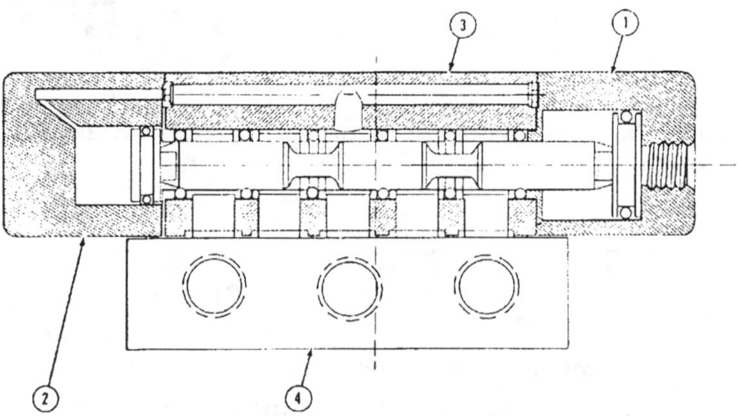

Figure 14.24 2-position, external pilot-operated, 4-way valve, with a differential pilot return: (1) pilot operator, (2) differential pilot operator, (3) valve body, (4) sub-base. (Courtesy Perfecting Service Company)

RETURN OPERATORS

Operators such as spring return reverse spring and spring center move the valve back to its normal (rest) position. The springs can be arranged in many ways in the valves to accomplish the required action. They can also be used with most types of actuating operators.

DETENTS

Detents are pre-determined locations used to control the positioning of a valve. They are used in many instances to prevent inadvertent valve movement. In manually-operated valves, detents provide the machine operator the desired "feel" of the different valve positions (Figure 14.6). Detents are incorporated in valves in many ways and can be used to indicate various positions.

VALVE POSITIONS

Valve position is the point at which the flow-directing elements provide a specific flow condition in a valve.

TWO-POSITION

A two-position valve is a directional-control valve having two positions for the selection of two flow conditions (Figure 14.9, 14.21, 14.24). Any two-, three-, four-, or five-way valve may be of a two-position type, depending on its operators.

THREE-POSITION

A three-position valve is a directional-control valve having three positions for the selection of three flow conditions (Figure 14.8). Any three-, four-, or five-way valve may be of a three-position type, depending on its operators.

Various flow conditions may be accomplished with spool valves, particularly by variations in their design (Figure 14.25). In addition to the two spool variations illustrated, there are also available in the series shown a standard two-position spool and a three-position open-center spool with the inlet port blocked and the cylinder ports open to exhaust.

NORMALLY-OPEN

A normally-open valve is arranged so that flow from supply to-work-port occurs when the valve is in its normal (rest) position (Figures 14.22A,

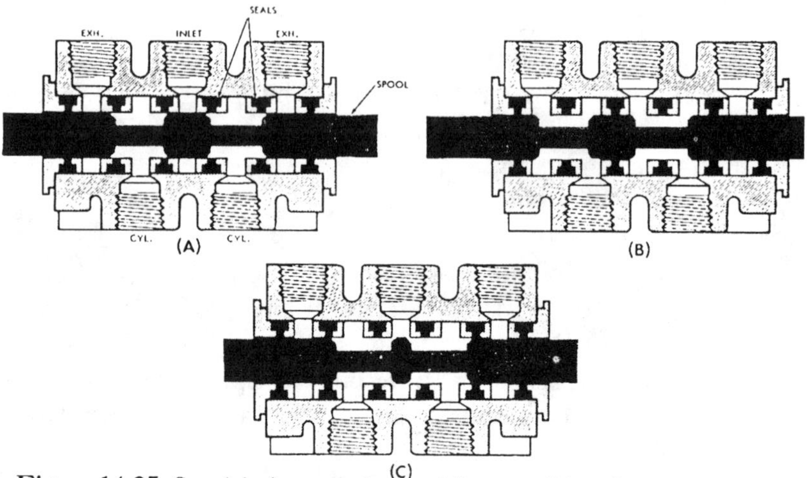

Figure 14.25 Spool design variations and flow conditions for a 4-way, 5-port spool valve: (A) all ports blocked in neutral position; (B) inlet port open to cylinder and cylinder port open to exhaust; and (C) exhaust ports blocked and cylinder ports open to inlet in neutral position. (Courtesy Automatic Valve Corporation)

14.26A, 14.26B). The two-way valve illustrated in Figure 14.26A may be used for on-off control of straight-through flow. The three-way valve shown in Figure 14.26B (normally-open) may be used to operate a single-acting cylinder. The inlet is normally open to cylinder and exhaust is blocked. When the valve is operated, the inlet is blocked and the cylinder port is open to exhaust. When the operating force is removed, the valve will automatically return to its normal (rest) position.

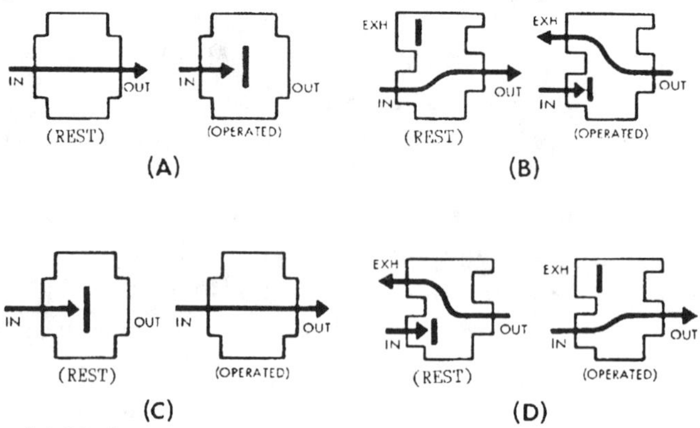

Figure 14.26 Schematic diagrams of 2-position, spring-return, basic valve designs: (A) 2-way in line valve, normally-open; (B) 3-way in line valve, normally-open; (C) 2-way in line valve normally-closed; (D) 3-way in line valve normally-closed. (Courtesy C. A. Norgren Company)

An important application of normally-open valves is in safety circuits. To prevent machine tables from being run off the machine, for example, a normally-open two-way valve would be placed so it would be actuated when the table reached a certain point in its travel, shutting off the table feed and stopping the table.

NORMALLY-CLOSED

A normally-closed valve is arranged so that when in its normal (rest) position, no flow takes place from the supply to the work-port (Figures 14.10, 14.22B, 14.26C, 14.26D). The two-way valve illustrated in Figure 14.26C may be used for on-off control of straight-through flow. The three-way valve of Figure 14.26D, normally-closed, may be used to operate a single-acting cylinder. The inlet is normally closed to the cylinder and the exhaust is open. When the valve is operated the inlet is opened to the cylinder and the cylinder port is blocked to exhaust. When the operating force is removed the valve will automatically return to its normally-closed (rest) position. This type valve is also used in safety circuits.

FLOW CONTROL VALVES

Flow-control valves "control" the speed of the cylinder piston stroke, allowing free, full flow in one direction and an adjustable (or restricted flow) in the other direction. The terms *"flow control"* and *"cylinder speed control"* are commonly used but are not accurate. The "control" of air flow is achieved *only* if the downstream pressure and the load resisting cylinder motion are constant. These valves may combine one or more of the design characteristics of directional-control valves or pressure control valves. Some of the more commonly used types are described in the following discussion.

FLOW-CONTROL, SPEED-CONTROL, AND BLEED-CONTROL VALVES

Exhaust flow on flow-control valves is controlled by a needle or long tapered valve stem which is externally adjustable. This can be pre-set for any flow from zero to full capacity. Graduated flow is pre-set by varying the air being exhausted. The speed of piston movement is controlled if its load is constant. For this reason flow-control valves are also mistakenly called *speed-control valves*. Review the data on this topic provided in Figure 4.1. Keep in mind that a pneumatic flow control valve does not have the flow-control capabilities of a hydraulic flow-control valve.

Figure 14.27A illustrates a poppet design, line-type "speed control" air valve which provides for full air flow towards the cylinder and for restricted air flow away from the cylinder through the adjustable needle valve. Figure 14.27B shows a floating-ball check-type flow-control valve which provides for full flow through the ball orifice and for restricted flow through the adjustable needle orifice. Figure 14.27C is an example of a needle-type speed-control valve which has restricted flow through the needle in one direction only, exhausting to the atmosphere. Such "speed-control" valves are used on the exhaust ports only of directional control valves, thus require only one threaded end. This type valve is also designated as a bleed-control valve. *Caution* - it can damage some types of O-ring spool valves.

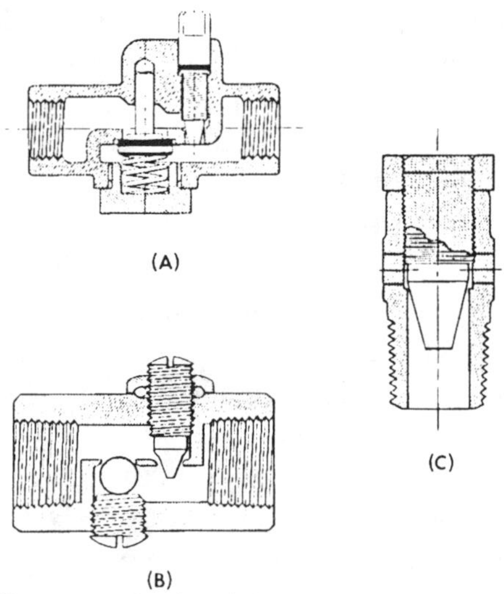

Figure 14.27 Flow-control valves (A) poppet design, line-type speed control valve; (B) floating ball-check type valve; and (C) exhaust-port, needle-type speed-control valve. (Courtesy Versa Products Company, Inc.)

A double-acting cylinder may be operated at different speeds in the power and return strokes of the piston by the use of two "flow-control" or "speed-control" valves. Each valve independently affects the piston velocity in one direction. Similar results may be obtained with the use of two exhaust "speed-control" valves on a five-port, four-way valve.

The volume of air on the exhaust side of the cylinder can be increased or decreased by locating the valves at various distances from the cylinder. This can be accomplished with a long stroke cylinder to obtain a decelerating effect. For some "speed-control" applications, it is best to install the "flow-control" valve close to the cylinder exhaust. In other instances, proximity to the cylinder is governed by accessibility to adjustments. Back-to-back positioning of the two valves will provide a measure of forward and return cylinder stroke control.

Where two double-acting cylinders are used and connected in parallel through the same directional-control valve, slow and fast piston speeds may be obtained through individually pre-set flow-control valves. They must be located between the cylinder and the common line to the exhaust of the valve.

For single-acting spring-return air cylinders, a surge tank or its equivalent may be used where delayed return of the piston is desired. The tank provides air volume which a connected "flow-control" valve can exhaust at a controlled rate to provide the time delay desired.

QUICK-EXHAUST VALVES

Quick-exhaust valves are used to accelerate the rate of action of cylinders, clutches, air brakes and other pneumatic devices by the rapid exhausting of air. These valves vent air pressure close to the operated device rather than through a control valve located at a further distance. Cycling or release time is thereby minimized. Fast dumping of air at the operated unit permits the use of smaller valves and piping.

Quick-exhaust valves are especially applicable to single- and double-acting cylinders. They may be used as a two-way directional-check valve in applications which require both low and high pressure air on the forward stroke of a cylinder. Faster action of the return stroke of an air cylinder is accomplished by attaching a quick-exhaust valve at the cylinder's air port.

An example of a quick-exhaust valve is shown in Figure 14.28. In the non-operating position (Figure 14.28A) the valve diaphragm rests on the exhaust port. When the control valve is opened, air pressure seals the exhaust port and the diaphragm is deflected. This permits air to pass through the cylinder port, as shown in Figure 14.28B. When the inlet pressure is relieved by the actuating valve in the control circuit, the build-up of the cylinder return pressure closes the inlet port. With the diaphragm released from the exhaust port, instantaneous release of air occurs, as indicated in Figure 14.28C.

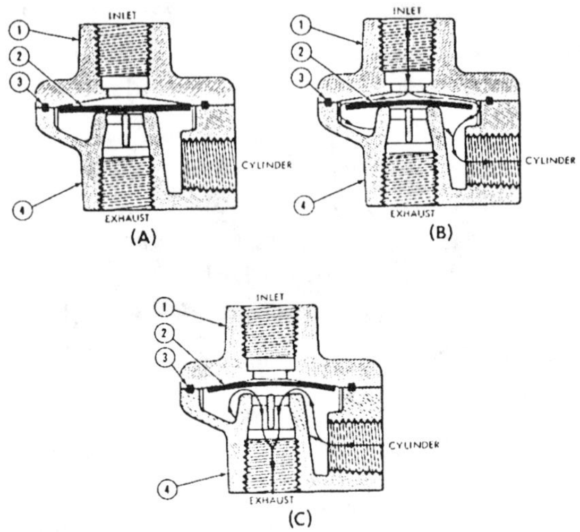

Figure 14.28 Cross section views of a quick-exhaust valve: (A) non-operating position: (1) bonnet, (2) diaphragm, (3) ring seal, (4) body; (B) exhaust port closed with air flow to cylinder; and (C) exhaust port open and cylinder exhausting. (Courtesy Auto-Ponents, Inc.)

CHECK VALVES

Check valves permit flow in only one direction (Figure 14.29). The check valve structure may use a poppet, ball or comparable closure as well as the swing gate-type shown. A simple check valve can be structured with an O-ring in a wide groove with a series of holes in the bottom of the groove. The assembly is installed in a suitable housing. Air can

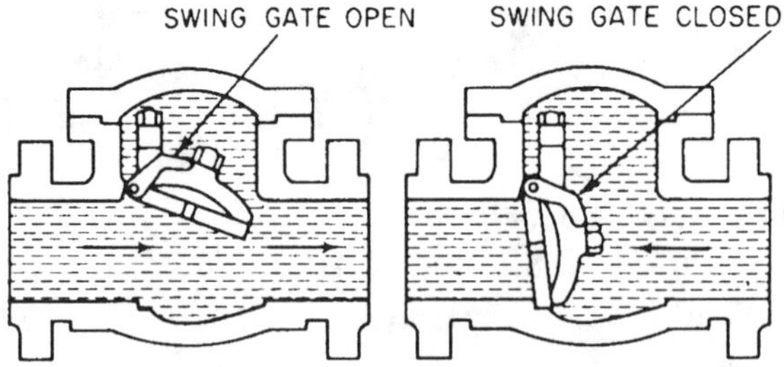

Figure 14.29 Swing check valve, commonly used by plumbers, may be found in pneumatic systems.

displace the O-ring as the pressurized air passes through the holes in the bottom of the groove past the sides of the O-ring and to the housing which is provided with an outlet port. Air flow in the opposite direction is stopped as the O-ring covers the holes in the bottom of the groove. Another simple check valve device for pneumatic service is called the duck bill check. A round elastomeric tube is molded with a flattened section. The pressurized air can expand the flattened section and pass through. Pressurized air in the opposite direction will close the flattened section, preventing a return flow. The elastomeric duck bill assembly is also placed in a suitable housing with the needed connections.

CUTOUT VALVES

Cutout valves stop air flow after a given amount of air has passed through the valve. This prevents the start of another cycle until the air upstream and downstream of the valve is exhausted.

SELECTOR VALVES

Selector valves function primarily to selectively interconnect two or more valve ports.

SEQUENCE VALVES

Sequence valves control the pre-determined sequence of circuit components operation.

SHUTTLE VALVES

Flow from B to C or from C to B can be accomplished in the shuttle valve of Figure 14.30 as long as pressure at A is at a lesser value. A higher pressure at A will urge the shuttle towards port B, and flow can then pass from A to C or C to A.

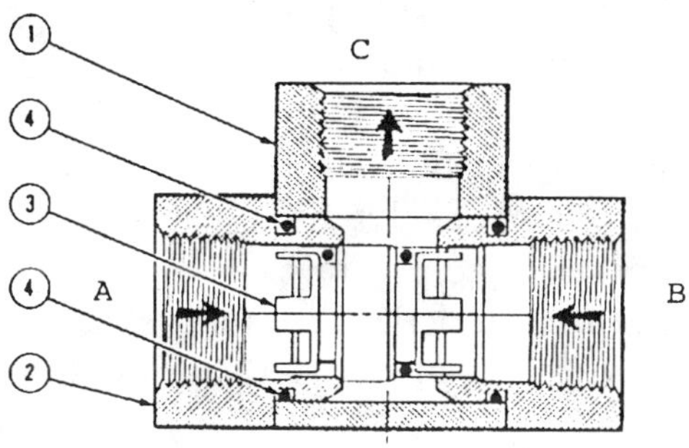

Figure 14.30 Shuttle valve: (1) body, (2) cap, (3) shuttle, (4) O-ring. (Courtesy Versa Products Company, Inc.)

SUMMARY

Pilot valves are valves used to control the operators of other valves. A pilot valve is operated by any means that supplies or exhausts air to actuate another valve. Pilot-operated valves are valves that are operated by air pressure controlled from another point.

Sequence valves perform the same function as pilot valves, but they are so connected in the circuit as to have interlock features.

Most all of the directional-control valves, depending on their use in the circuit, may be classed as sequence or pilot valves. Many of the valve nomenclatures refer to the operational function of the valve and not to the valve design.

FOR ZERO DOWNTIME

- Maintain cleanliness appropriate to the needs of the valve and actuator.
- Provide lubrication as specified by the manufacturer.
- *Do not over lubricate.*
- Install valves so that they are protected from mechanical damage and temperature extremes.
- Do not support lines with the valve, inducing stresses in the valve structure.
- *Do not tie down or defeat operator safety devices.*

REVIEW QUESTIONS

1. Name the three major types of pneumatic valves and describe their function.
2. What are the reasons for choosing a diaphragm-type regulator?
3. What are the advantages of a piston-type regulator?
4. What are the three most commonly used types of directional-control valves?
5. Describe the function of a bleed-pilot operator.
6. What is the purpose of a detent? What type of valves would incorporate a detent mechanism?
7. What is the primary function of a quick exhaust valve? How will it affect the size and operation of the primary directional control valve?
8. What types of valves are actuated by a solenoid?

9. Can a meter-out pneumatic flow control valve provide repetitively accurate rate of actuator movement?
10. What is the function of a cutout valve in a pneumatic circuit?

15

MANIFOLDS AND SUB-BASES

MOUNTING METHODS AND ARRANGEMENTS

Because of the almost limitless variety of usage, space requirements and functions for valves, there is no common method for their mounting. In the following general methods described, there are many arrangement variations. Those described and illustrated are suggestive of the more common types, and do not infer limitations with respect to other possible designs and arrangement combinations.

DIRECT-MOUNTING

A direct-mounted valve is one in which the ports are an integral part of the valve body and the valve operating mechanism, as illustrated in Figure 15.1. In some designs, auxiliary mounting holes and/or brackets are incorporated in the valve body or base for mounting on equipment or other convenient bases, as shown in Figures 15.2 and 15.3.

Direct-mounting has the disadvantage that the piping must be disconnected if and when the valve must be removed for repair or replacement. Direct-mounted valves have the usual advantage of lower first cost, which may in some cases outweigh the disadvantage of time involved in removing the valve from the piping.

Figure 15.1 Typical line-mounted pneumatic two-way valves. (Courtesy of Honeywell Inc., Skinner Valve Division)

Figure 15.2 Brackets attached for valve and line support. (Courtesy of KIP, Incorporated)

SUB-BASE MOUNTING

A sub-base mounted valve is one in which the ports are an integral part of the valve base or plate to which the valve body and valve operator(s) are mounted with threaded fasteners. Seals are used between the assembled units.

Figure 15.3 Direct-mounted valves. The ports are an integral part of the valve body and the valve operating mechanism. Mounting holes are provided as an optional mounting facility. (Courtesy of MAC Valves, Inc.)

Sub-bases and connections may be of various configurations. For solenoid-operated valves, they may include the electrical connections and plug-in terminals for the solenoid operators. Sub-base mounting has the advantage of permitting the user to remove or replace the complete valve and operator(s) when required without disconnecting or disturbing the piping. The incorporation of the electrical connections for the solenoid operator(s) in the sub-base has similar advantages for the electrical circuitry. Some sub-base designs have been standardized at the interface between the valve and the base. Some automobile manufacturers have established proprietary standards for equipment to be furnished in their plants. The American National Standards Institute has also been active in a pneumatic directional valve sub-base interface. The International Standards Organization has also provided standards for world-wide acceptance. There have been differences between these standards (Figure 15.4). When replacing valves which have been in service it is important to make certain the sub-base configurations are compatible. Figure 15.5 illustrates a 5-port, 2-position valve designed with the ISO sub-plate configuration.

Sub-base mounting is of major importance in automated facilities where minimum downtime is demanded. The sub-base in some cases may be incorporated in the operating equipment, thus eliminating the need for complicated piping. Provisions for proper connections must be included in the equipment as, for example, in the mounting of a valve directly on a cylinder.

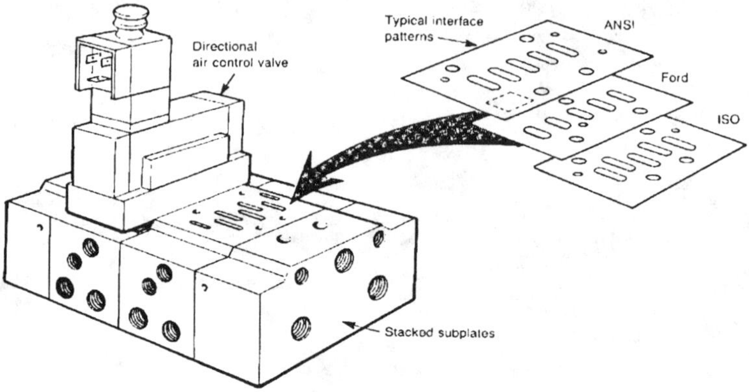

Figure 15.4 Standardization activities have been addressed by the American National Standards Institute in the United States (ANSI), automobile companies such as Ford Motor Company and the International Standards Organization (ISO) who publish standards which are used on a world-wide basis.

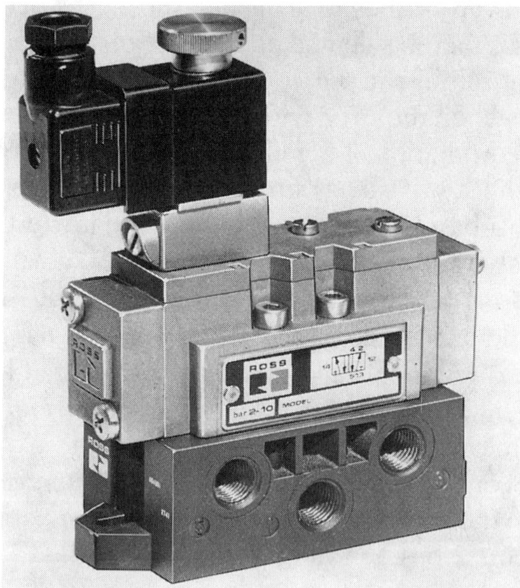

Figure 15.5 5-port, 4-way directional control valve designed to be mounted on a base with an ISO standard interface. (Courtesy of Ross Operating Valve Company)

The disadvantages of sub-base mounting are bulkiness and, in some cases, added valve cost.

Figure 15.6 Example of small size direct solenoid-operated, poppet type, 2- or 3-way valve for either normally-open or normally-closed service with 1-32 ports. A universal manual override is standard, which may be used either as a locking or non-locking override for machine set-up and trouble shooting. (Courtesy of Parker Hannifin Corporation)

STACKING AND MANIFOLD MOUNTING

Some valves are designed so they may be mounted adjacently or stacked as a group or unit. The individual valves can be held together by the use of tie rods or clamping devices. In those circumstances where the use of a number of manually-operated valves are involved, the location of all of the controls in one confined and accessible area is advantageous for maintenance, trouble-shooting, and operator accessibility. Manifold mounting is a combination of sub-base mounting and stacking. It has the advantages of both methods, as illustrated in Figure 15.6 and 15-7. Manifold and stacking methods simplify piping and in most cases, allow for common air inlets and common exhaust ports in the manifold.

SUMMARY

Line-mounted valves have the disadvantage that they must have piping removed for replacement. Subplate mounted valves can be replaced without disturbing major piping. Manifolds can reduce the need for external interconnecting lines.

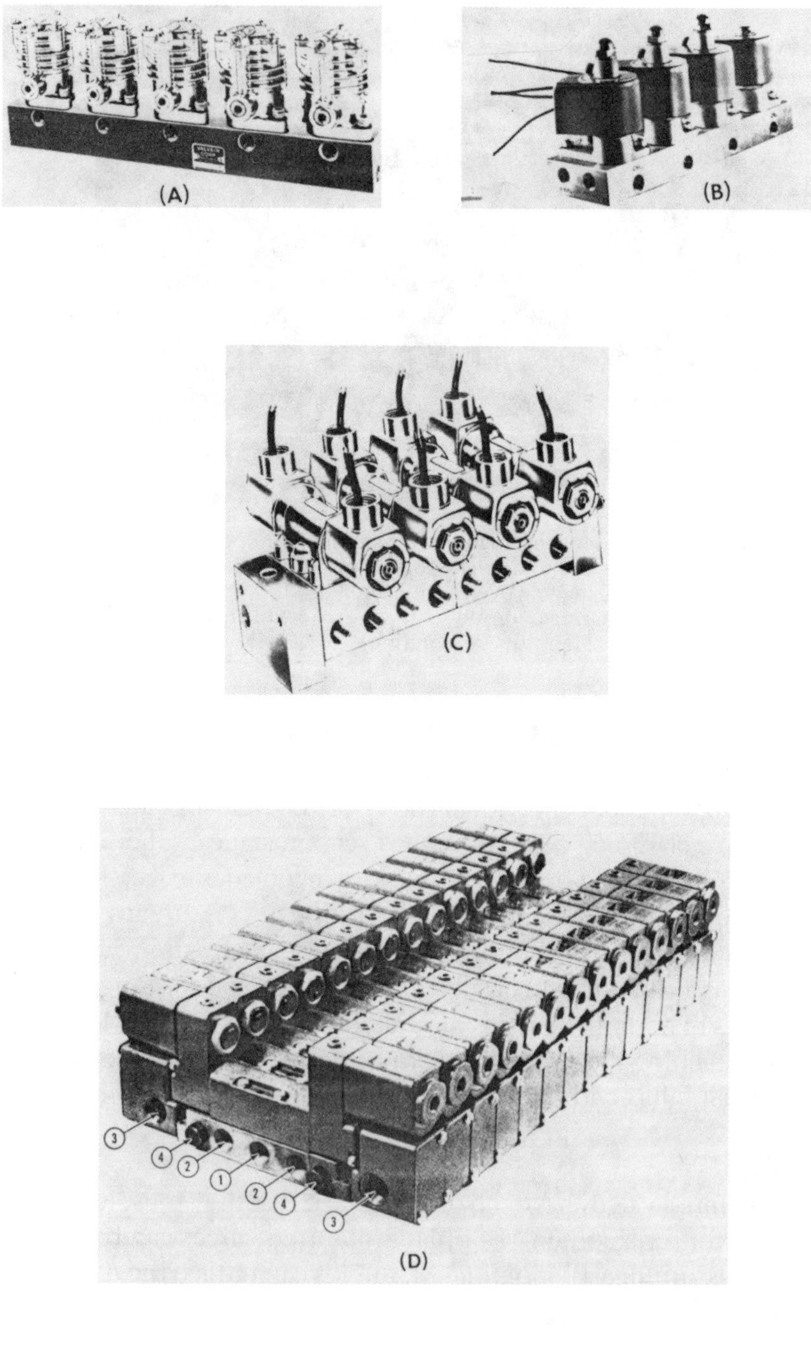
(A)
(B)
(C)
3
4
2
1
2
4
3
(D)

e

Figure 15.7 Examples of stacking and manifold mounting of valves: (A) reversible manifold base providing five solenoid-operated valves, sub-base mounted and stacked to form a manifold mounting, with common inlet and common exhaust ports; (Courtesy of Bellows-Valvair Division Parker Hannifin Corporation) (B) assembly of four solenoid-operated valves mounted on a single manifold, with a common inlet; (Courtesy of Automatic Switch Company) (C) four 4-way double solenoid valves, sub-base mounted and stacked to form a manifold mounting, with common inlet and common exhaust ports; (Courtesy of A. Schrader's Son) and (D) fourteen 4-way double-solenoid valves, sub-base mounted with plug-in solenoid, and stacked to form a manifold mounting, with common inlet, common exhaust and common electric conduit ports on both ends of the stacked unit: (1) inlet port, (2) exhaust ports, (3) electric conduit ports, (4) tie bolts. (Courtesy Perfecting Service Company). (E) Valve bank with integral pressure regulator. (Courtesy of MAC Valves, Inc.)

REVIEW QUESTIONS

1. What is the disadvantage of direct mounted valves?
2. What are the advantages of a sub-base mounted valve?
3. Can electrical connections be incorporated in a sub-base mounted valve assembly?
4. What are the disadvantages of sub-base mounted valve assemblies?

5. Name three standards that have been established for a sub-base mounting.
6. What is the purpose of stacking and manifold mounting?
7. Why is it important to tighten mounting bolts on sub-base mounted valves in a uniform pattern?
8. What type of mounting offers greatest initial cost savings?
9. What type of mounting offers greatest convenience to the operator and maintenance personnel?
10. Which type of mounting offers the greatest possibility to minimize external interconnecting lines?

16

LOGIC CIRCUITRY

Miniature air valves used only for decision control are often called logic valves, or just pneumatic logic. These small valves do no work in the typical force and motion context. But they do transmit YES or NO signals in the control circuit, just as electrical relays do in a switching circuit or transistors in a micro-processor. The final output pressure signal becomes pilot air or actuating input to a power valve or another physically small working device.

Air logic fills the gap in control hardware where the system is too complex for ordinary bulky control valves, yet not complex enough to warrant the addition of electronic microprocessor control. Also, air logic works well where only air is allowed in the process, for safety or other reasons.

The logic valves are often mounted in their own protective cabinet, where they receive signals from limit valves, push buttons, or special sensors driven by changes in pressure, proximity, flow, force, position and temperature.

The designs for air logic elements are as numerous as the manufacturers making them. Spools and poppet-type devices predominate, but many other concepts have worked - including fluidic devices with no moving parts.

ANSI Standard B.93.38-1976 (NFPA Standard T3.28.9-1973) is a method for diagramming moving parts fluid control (air logic) and contains definitions, symbols and circuits for all devices in common use. The designer or service personnel responsible for air logic devices should become acquainted with these standards and the technology involved with pneumatic fluid logic.

FLUIDICS

A lot of effort was made some years ago to create pneumatic logic without moving parts, based upon the invention of the fluid amplifier. Portions of this fluidic technology survived. However, the goal of replacing electric and pneumatic logic with fluid amplifiers never materialized, largely because of the economics of the electronic technology as it developed reliable, low cost devices.

The theory is that a stream of air or any fluid can be deflected or disturbed dynamically with a much weaker jet of fluid (the control), thus achieving "amplification" just as a transistor does. Figure 16.1 illustrates an example of fluidic amplification control.

Complete miniature logic and control systems were created out of all kinds of materials, including ceramics. They have been and continue to be successfully operated in a multitude of hazardous environments, and at extreme temperatures. Fluid amplifiers can stand enormous physical abuse because there are no moving parts to worry about.

Systems are in difficult applications such as jet engines, where they not only perform logic functions but also sense flow, temperature and pressure. Others are in use controlling ventilation systems, respiratory devices and nuclear power plants. Medical devices often employ fluidic principles in the operation of life-sustaining mechanisms.

One capability of fluidics has carried over into industrial controls, that of sensing as shown in Figure 16.2. A small sensing jet is disturbed when an object comes close. The resulting pressure build-up provides the desired signal to trigger an appropriate action in the machine cycle.

THE MICROPROCESSOR

Pneumatic power and control is quickly adapting to the electronic microprocessor. It was a long time coming because air itself is a versatile control medium, even without electronics.

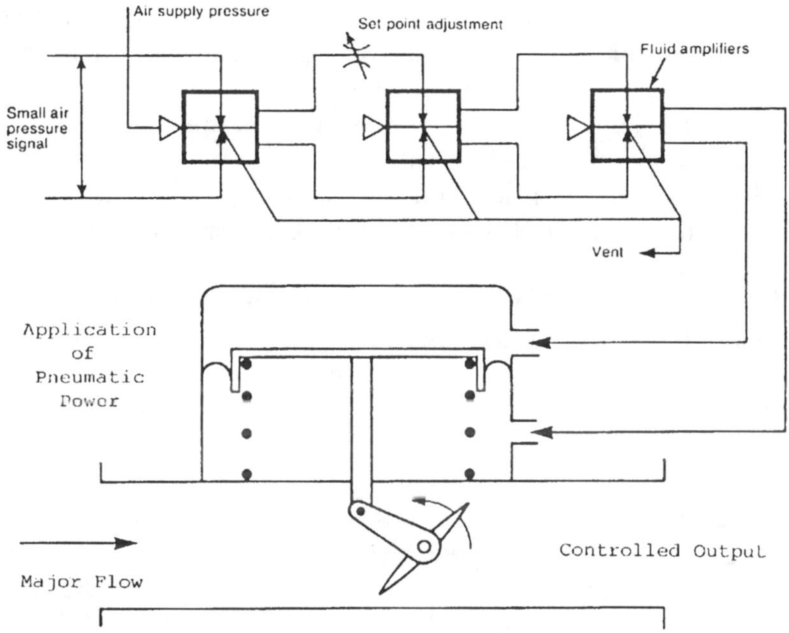

Figure 16.1 Fluidic control of a major flow line.

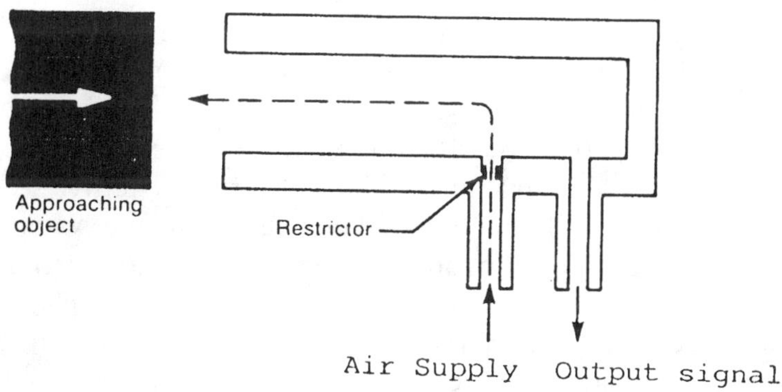

Figure 16.2 Fluidic sensing provides a signal to modify a machine cycle when an object approaches the fluidic sensor. No output signal is available when there is no restriction to air supply beyond the restrictor. As an object restricts the sensing port of the fluidic sensor, supply air is diverted to the output port and an output signal is created.

Microprocessors tell valves what to do with a precision and speed that cannot be achieved by air alone. Microprocessors help analyze, instruct, set, pulse, modulate and display.

As an example: air is highly compressible and cannot positively position a cylinder. But with position feedback monitored by a microprocessor control, an air cylinder becomes a precision device. The same is true for air motors. Pneumatic robots can respond as smoothly as electric robots, and usually at less cost.

The lowly on-off solenoid valve becomes an accurate pressure and flow controller when it is pulsed in a pattern pre-determined by the microprocessor. Or, if only an adjustable time delay is required, a modest number of solid-state electronic components integral to the coil can provide this function.

Proportional solenoid valves also have become useful accurately controlling flow by means of minute variations in current or voltage. They are giving electropneumatic servovalves competition.

Sensors integrate the electropneumatic controls: Hall-effect, magnetostrictive, magnetoresistive, reed switches, pressure switches and limit switches of all types.

MULTI-TALENTED DESIGN AND SERVICE

Pneumatic logic, fluidics, and electropneumatic systems require a knowledge of force and motion control by all known means, a knowledge of each of the disciplines and the ability to understand all of the symbology involved in each of the basic disciplines. Complete isolated pneumatic systems and complete isolated electronic/electrical systems are encountered less frequently. The integrated electro/pneumatic system serves industry in a wide spectrum of power transmission applications. The exceptions may be those associated with special hazards that preclude the use of electricity where total pneumatic control is essential. The other exception may be where a central pneumatic power supply is not readily available where a totally electric system may be economically justified.

SUMMARY

Air logic serves well where air only is allowed in the process, for safety or other reasons. Logic valves transmit YES and NO signals in the pneumatic control circuit. The final output pressure signal becomes input to a power valve or other working device.

ANSI Standard B.93.38-1976 (NFPA Standard T3.28.9-1973) is a method for diagramming moving parts fluid control (air logic) and contains definitions, symbols and circuits for all devices in common use. The theory associated with a fluidic device is that a stream of air or any fluid can be deflected or disturbed dynamically with a much weaker jet of fluid (the control), thus achieving "amplification" just as a transistor does.

Fluid amplifiers can stand enormous physical abuse because there are no moving parts subject to usual mechanical failure. Pneumatic power and control is quickly adapting to the electronic microprocessor. It was a long time coming because air itself is a versatile control medium, even without electronics.

The integrated electro/pneumatic system serves industry in a wide spectrum of power transmission applications. The exceptions may be those associated with special hazards that preclude the use of electricity where total pneumatic control is essential.

REVIEW QUESTIONS

1. What type of signals are provided for logic valves?
2. What organizations provide design and usage guidance with standards for air logic systems?
3. What type of signal systems function with no moving parts?
4. What is the objective of a sensing jet? What is the functional principal?
5. What type of hazards could preclude the use of electrical systems?
6. What is the basic theory of Fluidic devices?
7. Name five sensors used to integrate the electropneumatic controls.
8. How are microprocessors used in pneumatic circuits?
9. Why can fluid amplifiers stand enormous physical abuse?
10. Are there established standards for air logic elements?

17

SAFETY CONSIDERATIONS

1. TRAINING THE OPERATOR

The content of the training program will depend upon the machines to be used and will be different for operators, maintenance and set-up personnel. The machine designer will have additional responsibilities providing data and guidance in the development of training programs. The machine designer will need to be cognizant of the areas where training programs can make the difference between a safe, productive machine and a marginal or even dangerous machine.

The responsibility of the typical machine operator may be different because of the supervisory levels in the work place. Major manufacturing operations may have a sophisticated supervisory structure where the operator will be trained to recognize proper machine operation as it affects the product and rate of production. The operator can call for assistance if he or she questions the machine function and/or product quality. A set-up person may be available for tooling-related functions. A service person may be available to fill lubricators, clean filters and see to the mechanical health of the machine. Maintenance personnel may be available to make repairs. Manufacturing plants with less complicated staffing may combine the functional responsibilities of the various personnel. In some shops the operator must be acquainted with every basic motion of the machine he or she will control.

A qualified individual should be available to provide the needed instructions appropriate to the staffing structure in the manufacturing area. In some plants an operating manual may be the only source of information.

Therefore, in some shops, and depending upon the complexity of the process and equipment, the operator may have responsibilities which include that of calibrating the basic machine functions. He or she may be responsible for checking that quick-connect couplings are secure, be aware that quick connects larger than $^1/_2$-inch NPT may require special care in connecting and disconnecting; that hoses will not be snagged by a machine member; that air exhaust will not create a cloud of dust or otherwise contaminate the area for breathing; that lubricating devices are functional without an oil-fogged atmosphere around the machine, and that many other important machine mechanisms are in proper working order.

An operator may not need to be acquainted with pneumatic circuit drawings or basic pneumatic technology. But an operator *will* need to be acquainted with start and stop and emergency stop devices, safe operating pressure levels, safety interlocks, overload protection devices and overheat or exhaust icing indications.

Acquiring the dexterity to provide manual controls for some devices may come from dry run practice or relatively slow movements during the learning period. Most other training functions will be learned from the instructor or by study of the machine operations manual.

2. TRAINING SERVICE PERSONNEL

Service personnel will need to have acquired all the knowledge needed by the operators less possible manual skills and dexterity in actual control of certain machine functions. In addition, service personnel must be acquainted with mechanical drawings, pneumatic, electrical and electronic symbols, certain computer software and associated circuit drawings. Simple maintenance functions such as filter change or cleaning, lubrication requirements, and pressure level monitoring become an integral part of the service personnel tasks.

Signs of over-speed, overheat or incorrect mechanical alignment should be quickly noted by service personnel. Excessive sound level or changed sound should alert both the operator and service personnel to potential machine malfunction. Corrective measures are essential *before* the machine is damaged.

Service personnel *must read and understand* applicable maintenance manuals, know the basic machine operating cycles and be acquainted with the sound and pressure profile. Service personnel must be certain that conducting or non-conducting hoses are properly replaced to provide proper grounding or electrical isolation as needed.

3. TRAINING MAINTENANCE PERSONNEL

Maintenance personnel make necessary repairs to machinery. Their training will include that of the service personnel plus an understanding of the machine tooling and associated components so they can make complex adjustments, replace malfunctioning components and complete major repair functions. Well structured training is also available from distributors of fluid power equipment and components, from vocational technical schools, from fluid power component or system manufacturers and local chapters of the Fluid Power Society.

4. THE MACHINE DESIGNER'S RESPONSIBILITY

Obviously the designer of a machine, or one of the devices within a machine structure, must be fully acquainted with the end use function and capabilities of the machine. The machine designer must recognize the contribution of each machine member and the associated force and motion components, including both pneumatic power transmission and associated control mechanisms.

The mechanical and electrical/electronic functions are generally well documented with adequate training facilities available in most academic areas. Knowledge of the force and motion capabilities and the unique advantages of pneumatic power transmission are not as thoroughly documented, and adequate training facilities may not be readily available.

Many schools and technical institutes, however, have excellent educational programs in which pneumatic power transmission elements and systems are well documented and skilled instructors available. Appropriate laboratory facilities are usually provided so that the "hands on" knowledge gained in the laboratory enhances the educational process and the value of the instructor's expertise. Research is underway in some universities involving pneumatic power transmission systems and the associated motion and control components.

Many well qualified proprietary training programs are available. Most proprietary training programs provide an exposure to the basic sciences of pneumatic power transmission systems followed by specialized training relating to the proprietary system and equipment. Proprietary training can be a valuable adjunct to formal academic programs.

The machine designer has a significant obligation to position all machine members and control devices so as to pose minimum potential for danger to the operator, bystanders or machine in the event of a malfunction. If a potential for danger exists to the operator, nearby personnel or the machine itself it is mandatory that appropriate guards, shields or other protective devices be provided *before* the machine is placed into service.

The operating manual for the machine *must include all safety instructions* associated with set-up, operating, maintenance and repair procedures for the total machine and the integrated componentry. A pressure level and flow profile may be essential. The pressure level profile will indicate the pressure level at each machine operation. This ensures adequate pressure and flow to accomplish the task and ensure that excessive pressure does not damage the product or the machine.

Certain machines may require specialized safety devices such as two-hand controls for press operations to make certain the operator's hands are not in the danger area of a press when the die is in the closing mode. This function may be expanded to four or more hand controls when two or more operators are involved.

5. THE INSTALLERS RESPONSIBILITY

Power cut-off and/or emergency stops may be the joint responsibility of the designer, installer, plant engineer and setup personnel. Positive lock-out of electric and pneumatic power sources is essential when machine elements are being repaired, serviced or maintained. Exhausting all pressurized air is usually a *must.* Lock-out exhaust valves are readily available (Figure 17.1). Additional lock-out valves can be incorporated in the circuit to shut off and exhaust various segments. This means portions of the circuit can be available for maintenance service without interference with the circuit's other parts.

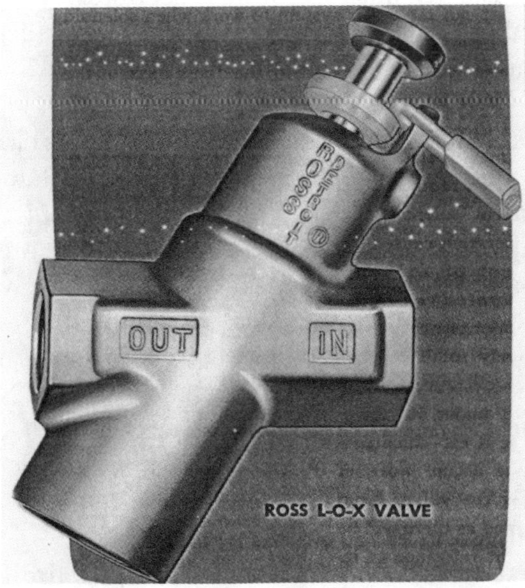

Figure 17.1 Lock-out exhaust valve. Locks out – with a padlock – air that operates a machine, as a safeguard when making machine repairs or leaving a machine unattended. The lockout valve is intended for installation in the supply line leading to the machine, upstream of any operating valve. When the valve is actuated, supply air to the machine is shut off and air downstream is exhausted. (Courtesy of Ross Operating Valve Company)

When a closed-center valve is used in a circuit, compressed air can be trapped downstream from it. Then, additional lock-out valves should be installed on each outlet line of the closed center valves. Lock-out valves may be detented in the open position. Most are designed for very low pressure drop.

The installer of the machine must provide adequate air supply and protection for flexible lines and moving machine members. Special care must be used if conventional quick connects larger than $^1/_2$-inch NPT are to be used in the pneumatic system.

Air exhaust must be directed so not to create a hazard for the operator or personnel in the area adjacent to the machine.

MINIMUM TRAINING REQUIREMENTS

OPERATOR

Know basic machine operating procedures. Alert to warnings of unusual or different sound levels. Watch product quality which may vary with the start of a malfunction.

SERVICE PERSONNEL

Understand machine circuit. Keep lubricators clean and filled to the appropriate level. Replace or clean filter elements. Watch for loose or damaged lines and repair as soon as practical. Check periodically for unusual or different sound level, pressure and flow levels, and product quality. Check and replace worn hoses with exact replacement, either electrical conducting or non-conducting type. Ensure quick connects are clean and safe. Use caution with conventional quick connects above $^1/_2$-inch NPT size. Some major users suggest that quick connects above $^1/_2$-inch NPT pipe sizes be reviewed by the safety department prior to installation on their machines or in their facilities.

SET-UP PERSONNEL

Understand and install tooling. Adjust limit or proximity switches for proper product tolerances. Makes pressure and flow adjustments appropriate for specific tooling, ensuring the pressure and flow are acceptable even in situations where only a few machines are operating. Such machines may have unusually high air pressure and flow.

MAINTENANCE PERSONNEL

Make repairs to malfunctioning machines. Replace faulty components. Make major adjustments. Schedule maintenance actions to be taken as a result of periodic inspection.

MACHINE DESIGNER

Prepares circuit and mechanical drawings for machine builder. Establishes training program needs. Determines service needs. Establishes basic safety considerations. Reviews design with machine builder to make sure components are located in a safe, accessible location.

INSTALLATION CREW

Install machine so that it is safe and can be serviced with minimum danger to machine or service personnel. Quick couplers larger than $^1/_2$-inch NPT generally need special handling or control devices to avoid injury to personnel and/or machinery, *especially* if they are used at pressures exceeding 60 psi. Alert all personnel working near the machine or handling the hose as to the safe and practical handling of large quick couplers. Higher pressures require extra care. Provide adequate protection for flexible lines. Determine that correct hose is installed. Make certain whether it should be electrically conducting or non-conducting as appropriate to the circuit and machine requirements. Make certain exhaust air is directed to a safe area without contamination of the atmosphere.

SUMMARY

The machine designer must know about the areas where training programs can make the difference between a safe, productive machine and one which is marginal or even dangerous. An operator must be acquainted with normal start, stop and emergency stop devices, safe operating pressure levels, safety interlocks, overload protection devices and overheat or exhaust icing indications. Acquiring the dexterity to provide manual controls for some devices may be by dry run practice or relatively slow movements during the learning period. Most other training functions will be learned from the instructor or by study of the machine operations manual.

Service personnel must have all the knowledge needed by the operators, less possibly manual skills and dexterity in actual control of certain machine functions. Signs of over-speed, overheat or incorrect mechanical alignment should be quickly noted by service personnel. Excessive or changing sound level should alert both the operator and service personnel to potential machine malfunction. Many well-qualified proprietary training programs are available. Most provide an exposure to the basic sciences of pneumatic power transmission systems followed by specialized training relating to the proprietary system and equipment. Proprietary training can be a valuable adjunct to formal academic programs.

Certain machines may require specialized safety devices such as two-hand controls for press operations to make certain the operator's hands are not in the danger area of a press when the die is in the closing mode. This function may be expanded to four or more hand controls when two or more operators are involved.

Positive lock-out of electric and pneumatic power sources is essential when machine elements are being repaired, serviced or maintained. Exhausting all pressurized air is usually a must. Lock-out exhaust valves are readily available. Additional lock-out valves can be incorporated in the circuit to shut off and exhaust various segments. This means portions of the circuit can be available for maintenance service without interference with the circuit's other parts.

REVIEW QUESTIONS

1. What is the responsibility of the machine designer to ensure a safe working environment?
2. What is the responsibility of the operator to ensure a safe working environment?
3. What are the responsibilities of service personnel related to the machine functions and a safe working environment?
4. What is the purpose of two-hand controls?
5. What is the purpose of a lock-out valve?
6. Who establishes basic safety considerations for the pneumatic power transmission system?
7. What are the responsibilities of the installation crew relative to a new machine or new installation?
8. Who makes pressure and flow adjustments appropriate to specific tooling?
9. Who replaces faulty components and makes major machine adjustments?
10. Why would a safety department review the installation of quick-disconnect couplers in $^1/_2$-inch sizes and above and determine whether additional safety cutoff valves are needed?

INDEX

A. Schrader's Son 213, 214, 237
Absolute pressure 31, 36, 39, 112
Accumulator 40, 41, 44, 46, 86, 88, 125, 149
Aerosol coalescers 141
Aftercooler 149, 160, 161, 163, 164
Air brake 46, 225
Air compressor 6, 8, 47, 59, 96, 98, 101, 103, 107, 109, 111, 160, 167
Air dryers 145, 147, 149, 156
Air line lubricators 150, 153, 155
Air logic 239, 240, 243
Air-Mite, Inc. 169
Alkon Products Corporation 220
ANSI 13, 240, 243
Argon 31
Aristotle 27
ASME 13, 35
Atmospheric pressure 35, 36, 37, 39, 46, 48, 62, 100, 112, 115, 118, 124
Automatic drain valve 127
Automatic Switch Company 237
Automation U.S.A. 179, 187, 216
Automatic Valve Company 222
Auto-Ponents, Inc. 226

Back pressure 81, 167, 170
Bellows 5, 97, 117, 118, 121
Bellowfram, Incorporated 90
Bellows Division, Parker Hannifin Corporation 47, 207, 208, 219, 237
Bernoulli's Principle 43
Bleed-control 223, 224
Bleed-pilot operators 219, 228
Bourdon tube gauges 112
Boyle's Law 29, 36, 39, 40, 44
Boyle, Robert 29, 39, 42

Carbon monoxide 31
Carter Controls, Inc. 90
Cavendish, Henry 30
Celsius 30, 31, 42
Charles, J. A. C. 30
Charles's Law 31
Check valve 16, 17, 23, 28, 99, 100, 151, 152, 225, 226, 227
Compression ratio 81
Crawford Fitting Company 183, 184, 185, 186
Cup seal 99
Cushioning 80, 81, 96

Cutout valves 227
Cylinder 39, 40, 45, 46, 48, 62, 64, 65, 66, 67, 68, 69, 70, 71, 73, 74, 75, 76, 77, 78, 79, 80, 81, 82, 83, 84, 85, 86, 92, 96, 98, 99, 100, 103, 138

Desiccant 144, 146, 147, 148, 149, 157
Deliquescent desiccant dryer 146
Design News ix
Detents 207, 212, 221
Diesel 5, 23, 109
Differential-pilot operators 220
Direct-mounting 231
Direct-pilot operators 218
Directional control valves 206, 224
Disc valve 208
Dowel pin 76
Drain valve 127, 130
Dryers 130, 141, 145, 146, 147, 149, 156

F-R-L (Filter, lubricator, regulator) 156
Fahrenheit 33, 34, 35, 37, 38, 39, 160, 161, 163
Filter 6, 8, 18, 23, 83, 103, 109, 122, 124, 130, 133, 136, 137, 141, 142, 143, 144, 145, 147, 148, 156, 157, 160, 178, 180, 245, 247, 250
Filter-regulator-lubricator units 156
Firestone Industrial Products Co. 90
Fisher Governor Co. 204
Five-way valve 211, 221
Flow control 20, 25, 46, 49, 81, 84, 85, 86, 169, 223, 224, 229, 242
Fluid Power Society 247
Fluidic 239, 240, 242, 243
Ford Motor Company 138
Four-way valve 21, 85, 211, 224
FPDA ix
Free air 12, 35, 36, 37, 39, 44

Galileo 28
Gast Manufacturing Company 87
Gauge pressure 36, 38, 112
Gay-Lussac 42
Glossary of Terms for Fluid Power 198
Ground effect 6

Hilsch 165
Honeywell, Inc. Skinner Valve Division 232
Hose and tube fitting 189
Hovercraft 96
Hydro-Check 46

Icing 165, 166, 169, 170, 246, 251
Illinois Institute of Technology 165
Installation crew 251, 252
Instrument-pure air 147
Ingersoll-Rand Company 108, 161
Instrumentation 8, 145, 156
Intercooler 110, 160
ISO (International Standards Organization) 13, 233

JIC (Joint Industry Conference) 13, 76, 215

KIP, Incorporated 232
Krypton 31

Lavoisier 30
Lehigh Fluid Power 89
Lock-out exhaust valve 249, 252
Logansport Machine Company 207
Logic 11, 30, 103, 141, 202, 239, 240, 242, 243
Low-pressure pilot operator 218
Lubrication 13, 24, 47, 48, 96, 102, 105, 139, 151, 152, 153, 156, 157, 165, 166, 167, 169, 187, 228, 247
Lubrication dispersion 166
Lubricator 6, 8, 136, 137, 141, 150, 151, 152, 153, 155, 156, 157, 178, 188, 245, 250

MAC Valves, Inc. 211, 212, 216-218, 233, 237
Machine designer 245, 247, 248, 251, 252
Maintenance personnel 24, 26, 125, 136, 238, 245, 247, 250
Manual operator 212
Maxwell 166
Maxwell's Demon 166
Mechanical 4, 5, 13, 16, 24, 35, 46, 48, 98, 111, 124, 164, 176, 180, 194, 212, 213, 214, 215, 218, 219, 228, 243, 245, 246, 247, 251

Mercury 28, 29, 30, 35, 112
Methane 31
Micron 124, 142, 143, 148
Microprocessor 239, 240, 242, 243
Morris, C.W. Co. 168
Motor 4, 6, 12, 17, 18, 20, 23, 26, 59, 62, 86, 88, 96, 107, 124, 166, 167, 169, 198, 206, 210, 211, 242
Muffler 142, 165, 166, 167, 168, 169, 170

NCFP ix
Neon 31
NFPA (National Fluid Power Association) 171, 174, 199, 240, 243
Nitrous oxide 31
Norgren, C.A. Co. 200, 203, 210, 222
Normally closed 21, 22, 202, 223
Normally-open 21, 221, 222, 223
Numatics, Incorporated 215
Nycoil Company 176

Oil-free 103, 110
Oil-less 103, 104, 110
Oil-lubricated 103
Operating manual 24, 25, 246, 248
Operators 11, 26, 212, 213, 214, 215, 218, 219, 220, 221, 228, 233, 245, 246, 248, 251, 252
Origa Inc. 90
OSHA 125, 153

Papin 4
Parker Hannifin Company 81, 83, 95, 132, 154, 157, 209, 235
Pascal, Blaise 30, 42
Pelton Water Wheel 1
Penton/IPC Publishing 147, 163
Perfecting Service Company 190, 191, 192, 193, 204, 220
PHD, Incorporated 91-94
Phlogiston 30
Piab USA, Incorporated 115, 116, 117, 119, 120, 121, 122, 123
Pilot operators 218, 219, 220
Pilot-operated regulator 201
Pipe fittings 176, 177, 195
Piston or spool valve 206
Piston rod diameter 78
Pivot mount 71
Plug valve 208
Plunger valve 208
Pneumatic actuators 66, 88
Pneumatic drills 98
Pneumatic motor 17, 62, 86, 166, 169
Poppet valve 198, 210
Porous stone 142, 143
Pressure control valve 198, 223
Priestley 30

Quick-disconnect couplings 190, 191, 192, 195
Quick-exhaust valves 225

Ramsey, William 31
Refrigerant cooler 141
Refrigeration 145, 146, 162, 165
Regulator 6, 8, 22, 26, 127, 133, 136, 137, 141, 156, 178, 180, 197, 198, 200, 201, 202, 204, 205, 206, 228
Relief valve 13, 21, 22, 81, 85, 101, 125, 206
Reservoirs 8, 156
Return operator 213, 221
Robot 8, 88, 92, 121, 124, 242
Rod end joint 77, 78, 79
Rod seals 82
Rod wipers 83
Ross Operating Valve Company 12, 25, 234, 249
Rotary joints 194
Rutherford 30

Safety considerations 26, 245, 251, 252
Selector valves 227
Self-aligning 73, 75, 190
Separators 130, 133, 134, 145, 148
Sequence valves 227, 228
Service personnel 25, 156, 197, 240, 246, 247, 250, 251, 252
Set-up personnel 245, 250
Shear key 73, 74
Shear-action design 206
Shuttle valve 227
Sintered-bronze 142, 143
Solenoid operators 233
Southwest Industries 177
Spectroscopy 31
Speed control 80, 84, 86, 223, 224
Stacking and manifold mounting 235, 238

Stahl, George 30
Standard air 35, 36, 44
Stop tube 79, 96
Sub-base mounting 232, 233, 234, 235, 238
Swivel joints 194
Symbology 11, 13, 242
Synthetic lubricant 103, 109, 141

Tank drains 145, 148
Terminology 13, 14, 198
Thread protectors 186, 194
Three-way valve 210, 211, 216, 222, 223
Tol-O-Matic, Inc. 90
Torricelli, Evangelista 28
Torricelli's Experiment 31
Torricellian Barometer 30
Torricellian vacuum 43
Training 11, 26, 245, 246, 247, 248, 250, 251, 252
Trunnion 71, 73, 75
Tube cutting 183
Tube deburring 185
Tube fitting 16, 178, 182, 185, 189, 190
Tube support 180
Two-way valve 210, 211, 222, 223

Vacuum 8, 9, 17, 27, 28, 29, 43, 45, 46, 48, 55, 59, 62, 100, 111, 112, 113, 114, 115, 118, 120, 121, 122, 124, 167
Vacuum forming 9, 62, 118, 124
Valve operator 212, 232
Valve porting 210
Valve positions 221
van Helmont, Jean Baptista 33
Versa Products Company, Incorporated 224, 227
Vortex cooler 141

Water vapor 32, 33, 34, 35, 37, 38, 39, 127, 160, 161, 162
Westinghouse, George 46
Wilkerson Corporation 29, 31, 32, 34, 37, 38, 129, 139, 144, 146, 149
Windmill 1, 55
Wire mesh 142, 143

Xenon 31

Yeaple, Frank ix, x